AUDIO
SYSTEMS

AUDIO SYSTEMS

Clyde N. Herrick

San Jose City College

RESTON PUBLISHING COMPANY, INC.
Reston, Virginia 22090
A Prentice-Hall Company

Library of Congress Cataloging in Publication Data

Herrick, Clyde N
  Audio systems.

    1. Sound—Recording and reproducing.   2. Electronic
apparatus and appliances—Maintenance and repair.
I. Title.
TK7881.4.H47   621.38'0412   74—9696
ISBN 0–87909–049–9

© 1974 by
Reston Publishing Company, Inc.
A Prentice-Hall Company
Reston, Virginia 22090

10   9   8

Printed in the United States of America.

CONTENTS

PREFACE

Audio devices, circuits, and systems have evolved rapidly within the past several years. Quadriphonic sound has become an important activity in the high-fidelity field. The bipolar and field-effect transistor, the integrated circuit, the light-dependent resistor, and related devices have made obsolete the traditional audio building blocks. Intercommunication systems have become more sophisticated and are often combined with high-fidelity equipment and paging units. Public-address systems have also become more sophisticated, owing both to recognition of the Haas effect and to the demand for high-fidelity reproduction in concert halls and ballrooms. Improved microphone and speaker designs have aided the development of PA systems significantly.

Broadcasting studio and audio systems have also evolved rapidly. Improvement of field audio facilities is particularly marked, both from the standpoint of utility and operating convenience, as well as improved reliability and quality of performance. Commercial telephone systems have become highly sophisticated in recent years, and human operators are still employed only in PBX systems. Automation has been accelerated by introduction of solid-state devices and computer technology into commercial telephone switching systems. Theater sound systems provide greater realism and higher fidelity than in the past. Moreover, new techniques have been introduced that provide greater flexibility for actors, directors, conductors, and technicians in studio activities.

Electronic organ sound systems have kept pace with advances in other areas of the audio field. The modern electronic organ with its polyphonic characteristics is a far cry from first-generation melodic organs. As in various other audio fields, computer technology has made important contributions to modern organ design. Although less prominent than various other audio systems, carrier-current audio systems have undergone extensive development and are now a dominant aspect of telephone engineering practice. Simultaneously, power-line carrier communication reflects modern electronic advances, thereby becoming more efficient and utilitarian. In a more visible and dramatic area, new-music (electronic-music) audio systems have captured wide popular attention. Whether one likes or dislikes the new music, it is "growing up" rapidly and cannot be ignored. Finally, audio measurements have become increasingly sophisticated and make greater demands on the expertise of audio technicians than in the past.

To provide a broad foundation of understanding and to facilitate transfer of training by the student, the conceptual approach has been emphasized in this text. On the other hand, the "hardware" aspect of the discipline has not been entirely ignored, inasmuch as the tangible facets of audio systems make a legitimate contribution to overall perspective. Mathematics has been introduced only to the extent required to provide ample rigor at the introductory level. Prerequisite courses are basic electricity, electronics, and semiconductor technology. However, an otherwise qualified student can cope with this text successfully if he is taking a concurrent course in semiconductor technology.

Troubleshooting of audio circuits and systems is systematically covered in this text, inasmuch as the treatment is oriented toward both the vocational student and the college-preparatory student. Grateful acknowledgement is made to the manufacturers who have been credited in the text for their cooperation and their generosity in providing photographs, diagrams, and technical data. Acknowledgment is also made to the faculty of San Jose City College, who have provided numerous constructive criticisms. An author does not work in a vacuum, and, in a significant sense, this text represents a team effort. It is appropriate that this book be dedicated as a teaching tool to the instructors and students of our technical schools and junior colleges.

Clyde N. Herrick

1

HIGH-FIDELITY COMPONENT SYSTEMS

1.1   GENERAL CONSIDERATIONS

Most high-fidelity connoisseurs prefer component systems. For example, a chosen pair of speakers may be used with a preferred type of two-channel amplifier, plus a selected brand of record player, a desired design of AM-FM tuner, a chosen reel-to-reel tape deck, and/or an eight-track tape deck, or a favored cassette deck. On the other hand, high-fidelity stereo instruments are also available in unitized form and housed in elegant furniture cabinets. A stereo equipment *console*, as the unitized arrangement is termed, eliminates the necessity of planning a component system. Some consoles contain equipment that favorably compares with a first-rate component system. Other consoles have mediocre equipment housed in expensive furniture cabinets. A console contains a speaker in each end of the cabinet.

One type of stereo instrument, termed the *compact*, has sepa-

1

rate speakers, with a record turntable and stereo amplifier on the same base. The speakers are often mounted or placed about twelve feet apart. The main unit of a compact may contain an AM and/or FM tuner in addition to a record turntable. Another design of compact has a record changer mounted on top of the main unit, under a clear plastic cover. A compact is sometimes called a *modular* hi-fi unit. This usage of the term *modular* should not be confused with the printed-circuit modules that may be used in amplifiers. Amplifier modules are explained in greater detail subsequently.

A hi-fi buff often selects the speakers first. A matched pair of speakers, such as illustrated in Fig. 1–1, is employed in a stereo system. A pair of high-quality stereo speakers usually costs from $200 to $400. However, good-grade high-power speakers are more costly. Some hi-fi enthusiasts like to listen at comparatively loud levels and larger speakers are required to handle the additional audio power. A bookshelf-type speaker, as shown in Fig. 1–2, is adequate for moderate volume levels, but it will distort the sound output or burn out if driven at high power levels.

A hi-fi speaker cabinet contains more than one speaker unit. The largest speaker, called a *woofer*, is used to produce the low bass tones. The smallest speaker, called a *tweeter*, is used to reproduce the high treble tones. A speaker of intermediate size, often called a *squawker*, or mid-range speaker, is connected to the system so that it reproduces the midde range of tones between the low bass and the high treble tones. Some speaker cabinets contain a pair of middle-range speakers, one of which is larger than the other. In general the size of a speaker corresponds to the audio power that it can handle.

FIG. 1–1. A matched pair of high-fidelity speakers. (Courtesy of Electro-Voice Inc.)

FIG. 1–2.   A bookshelf type speaker. (Courtesy of Heath Co.)

A bass speaker is always the largest because the bass tones carry the greatest proportion of the audio power in most musical passages. The speakers and their associated electrical networks in a speaker cabinet, or enclosure, are referred to as a *speaker system*.

Many stereo amplifiers and various compact instruments provide jacks for plugging in stereo headphones. Some hi-fi buffs prefer the acoustics of headphones, while others prefer headphones for privacy. Stereo amplifiers are designed for very low distortion, for a rated audio-power output, and for various types of inputs. These inputs concern the desired program sources. As an illustration, an AM-FM tuner, a reel-type tape deck, and a cassette player each needs an appropriate *input jack*. An amplifier also provides *features* such as tone controls, loudness-type volume control, terminals for additional speakers, filters, stereo balance control, and various jacks as noted above. Hi-fi enthusiasts who make their own tape recordings require an amplifier that provides an appropriate stereo signal for a particular tape recorder. (See Fig. 1–3.)

Figure 1–4 shows a typical stereo amplifier chassis, and Fig. 1–5 illustrates a *receiver* consisting of a tuner and an amplifier. Few separate stereo tuners are being manufactured today; most tuners are incorporated with amplifiers. All stereo tuners have a multiplex decoder to provide stereo sound from stereo FM stations. Some tuners include a Dolby noise-reduction network. Note that hi-fi record changers are generally called *turntables*. An *automatic turntable* holds a number of discs and plays one side of each in

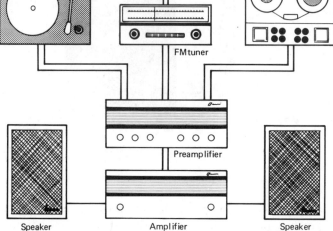

FIG. 1–3.   Components of a typical hi-fi stereo system.

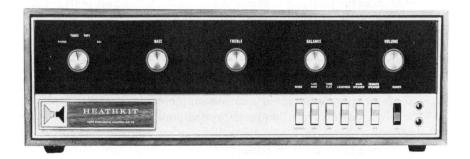

FIG. 1–4.   A stereo amplifier chassis. (Courtesy of Heath Co.)

sequence. The most sophisticated types of hi-fi record changers are called *transcription turntables*. Even the simplest types of record changers provide stereo reproduction, with left-channel and right-channel output signals.

A tape recorder provides both recording and playback facilities, whereas a tape player lacks recording facilities. A tape deck

FIG. 1–5.   A receiver, consisting of a tuner and an amplifier.

does not have an amplifier and must be used with an external amplifier and speaker system. Note that a tape deck may or may not provide recording facilities. Figure 1–6 illustrates a professional

FIG. 1–6.   A professional quality reel-to-reel tape deck. (Courtesy of Ampex)

quality reel-to-reel tape deck. Monophonic recording is accomplished with a single microphone (or audio signal source), whereas stereo recording requires a pair of microphones (or a stereo signal source). Audiophiles tend to prefer reel-to-reel machines over cartridge or cassette-type machines. Eight-track *cartridge* tape players have become popular with the general public because they are compact and comparatively simple to operate. The majority of eight-track cartridge tape machines are *player decks*. In other words, they lack recording facilities.

More recently, the *cassette* cartridge machine employing *chromium dioxide* tape has made inroads into the audiophile market. This type of machine provides hi-fi reproduction of cassette recordings that approaches the performance of high-quality reel-to-reel machines. Eight-track tape players are very widely used in automobiles. All these types provide stereo reproduction and many qualify as high-fidelity units. High-fidelity reproduction denotes a frequency response that is flat within $\pm$ 1 dB from 20 Hz to 20 kHz and a distortion level less than 1 percent at maximum power output. Note that hi-fi speakers generally have a frequency response that is flat within $\pm$ 6 dB from 20 Hz to 20 kHz.

1.2   AM TUNERS

All high-fidelity consoles provide an AM tuner, although very few of these are capable of hi-fi reproduction. Note that an AM tuner is a high-quality design of a conventional AM radio receiver without an audio amplifier and speaker. There are two basic reasons for the lack of hi-fi reproduction from an AM tuner. First, very few AM broadcast stations provide high-fidelity transmissions. There is seldom more than one hi-fi AM broadcast station in a metropolitan community. Second, since there are so few hi-fi AM broadcasts available, AM tuners are infrequently designed for hi-fi response. Tuners designed for conventional response have the typical selectivity curve depicted in Fig. 1–7. In turn, the audio-frequency response for a conventional AM tuner is as shown in Fig. 1–8. This response can be improved to some extent by means of treble boosting, but it cannot be brought up to high-fidelity standards. If an AM tuner is designed for hi-fi reproduction, a complaint of poor selectivity and adjacent-channel interference is likely to be brought up by the customer. Accordingly, some hi-fi AM tuners have been

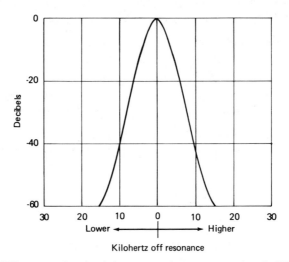

FIG. 1–7. A selectivity curve for a conventional AM broadcast receiver.

designed with a selectivity switch for choice of either wide-band or conventional response.

Figure 1–9 shows a block diagram for an AM tuner with adjustable bandwidth, and Fig. 1–10 depicts the schematic diagram for this tuner. Symptoms of malfunction include distorted reproduction, weak audio output, reception interference, tuning drift, intermittent operation, and no audio output. Most defects in AM tuners involve defective capacitors. Electrolytic capacitors, such as C28, have the highest failure rate. They may lose capacitance,

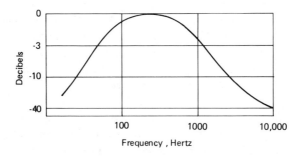

FIG. 1–8. Audio-frequency response for a conventional AM tuner.

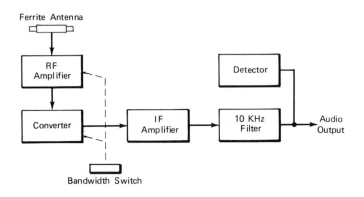

FIG. 1–9. Block diagram for an AM tuner with adjustable bandwidth.

become leaky, or develop a poor power factor. Fixed-paper capacitors are likely to eventually become leaky, and sometimes become open-circuited. Semiconductors are less likely to cause trouble, but should be checked if other component defects are not found. Occasionally, a resistor will drift seriously off-value or become intermittent. Sometimes, cold-soldered joints and cracks in printed-circuit wiring cause trouble symptoms. Note that alignment is always checked last, after troubleshooting has been completed, unless it is known that the set-owner has tampered with the alignment adjustments.

When troubleshooting an AM tuner, it is advisable to employ a signal generator. The generator can be used as a signal injector to check the operation of each stage working back from the detector. For example, if a modulated IF signal is injected at the left-hand end of CR1 in Fig. 1–10 and no signal output is obtained from the tuner we may conclude that the diode is probably either open-circuited or short-circuited. However, if the diode shows a normal front-to-back ratio in a resistance test, we may then suspect that C30 is open-circuited. Note that a short-circuit in C23 will also cause a no-output symptom. DC voltage measurements are usually the best guide to a defective component, after a faulty stage has been localized. Receiver service data specifies normal DC voltages throughout the receiver circuit. Abnormal or subnormal voltage values indicate that a defective component is associated with the test point.

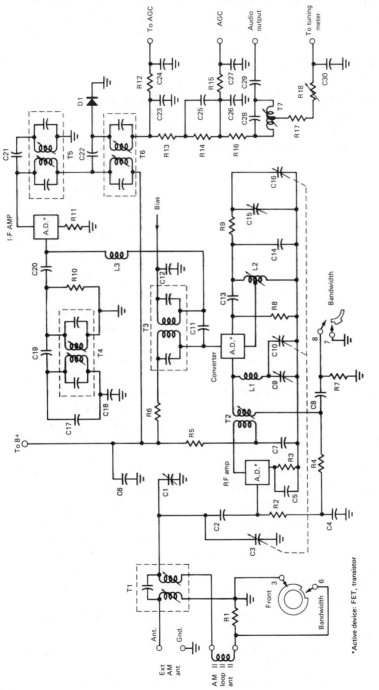

FIG. 1-10. Schematic diagram for adjustable-band-width AM tuner.

*Active device: FET, transistor

9

1.3   FM TUNERS

An FM tuner has inherent wide-band response and is capable of high-fidelity reproduction. As shown in the block diagram of Fig. 1–11, an FM tuner is basically an FM receiver, minus an audio amplifier and speaker(s). Numerous FM broadcasts are stereophonic, so that a stereo multiplex decoder can be used with the basic arrangement of Fig. 1–11 to obtain left and right stereo outputs for driving a pair of audio amplifiers and speakers. This process is detailed in Section 1.4. A schematic diagram for a typical FM tuner is shown in Fig. 1–12. This configuration is somewhat more elaborate than that depicted in the block diagram of Fig. 1–11 in that the IF section provides both 10.7-MHz FM IF transformers and 455-kHz AM IF transformers. Both FM and AM detectors are also provided. Thus, the IF section operates in combination with either the FM tuner, or with an AM converter section (not shown). This design is employed in consoles for simplification and production economy.

Symptoms of component defects include distorted audio output, noisy output, poor sensitivity, oscillator drop-out at high end of band, interference, intermittent operation, or no reception. Distortion is most likely to be caused by a defect in the FM detector circuit. For example, if CR2 (Fig. 1–12) becomes open-circuited, short-circuited, or develops a poor front-to-back ratio, the audio output will become distorted and attenuated. Noisy output is likely to be caused by capacitor defects, such as loss of capacitance in C23. Oscillator drop-out is generally caused by a marginal defect in the converter transistor, although a leaky capacitor in the converter stage can cause the same symptom. Interference is usually caused by misalignment, which results from an open capacitor in a tuned-circuit network. Intermittent operation can be caused by cold-soldered joints, breaks in printed-circuit conductors, or marginally open capacitors. A transistor occasionally becomes intermittent, although this is less likely than component intermittents. No reception can be caused by defective switches or by open or shorted capacitors.

Trouble localization is facilitated by the use of an FM signal generator. The generator can be used to inject a test signal step-by-step, working back from the FM detector. After a defective stage is localized, DC voltage measurements are generally most useful to

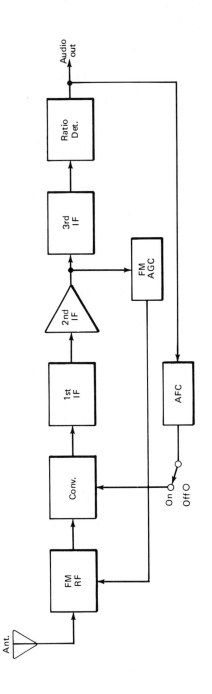

FIG. 1–11. Block diagram of a typical **FM** tuner.

11

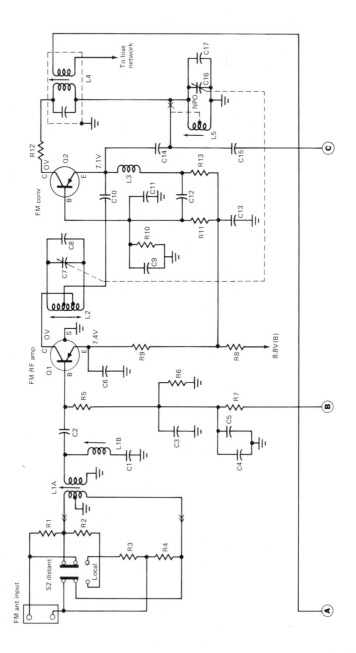

12

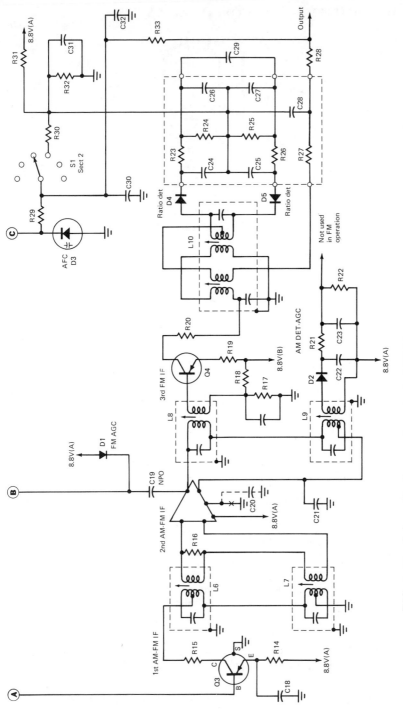

FIG. 1–12.   Schematic diagram for a typical **FM** tuner.

13

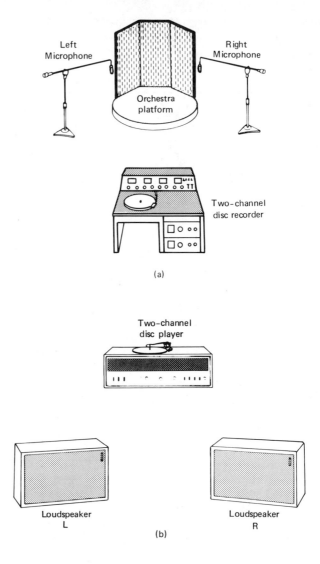

FIG. 1–13. Basic stereophonic sound system. (a) Recording of a disc. (b) Record playback.

pinpoint a defective component. Figure 1–12 is an example of a schematic diagram that specifies normal operating voltages. A component defect is often associated with abnormal or subnormal voltages in its associated circuit. Note also in Fig. 1–12 that various coil-winding resistances are specified. Resistance measurements often serve to supplement DC voltage measurements in verifying or negating a preliminary conclusion. Sometimes DC current measurements are also useful. For example, the integrated circuit used in the second IF stage (Fig. 1–12) has a rated current drain of 8 mA at terminal 8. If the IC is found to be drawing excessive or insufficient current, we would conclude that the IC is defective and needs to be replaced.

1.4   STEREO FM MULTIPLEX DECODERS

A stereophonic sound system employs two separate sound channels, which are shown in Fig. 1–13. Similarly, an FM stereo system has two effectively separate sound channels, which are depicted in Fig. 1–14. However, since both channels must be transmitted by the same electromagnetic waves, a technical expedient is required to maintain channel separation in an FM transmission. In other words, the R channel is transmitted in the usual manner, but the L channel is *encoded* into the transmitted signal, as will be explained. In turn, a *decoder* is required at the stereo FM receiver to recover the L audio information. If a decoder is not used, only the R audio information is reproduced. As shown in Fig. 1–15, FM broadcast channels have been allocated by the FCC for a bandwidth (deviation) of 150 kHz, or ±75 kHz. A guard band of 25 kHz is provided on either side of the channel, and transmission is forbidden in the guard band. Since the conventional FM signal (monophonic signal) occupies the entire 150-kHz channel, it is evidently necessary to employ a *multiplexing* arrangement to transmit a stereo signal.

Next, the multiplexing arrangement must provide *compatibility*. In other words, a stereo broadcast must provide separate L and R audio channels for a stereo receiver, and the same broadcast must provide normal reception by a monophonic receiver. Two microphones are employed in a stereo broadcast arrangement. As depicted in Fig. 1–16, a normal monophonic signal is produced when the L and R microphones are operated parallel to each other. This monophonic signal is called the $L+R$ signal. Multiplexing starts with a 38-kHz subcarrier signal, as shown in Fig. 1–17. The subcarrier frequency is out of the audible range; therefore reception of the $L+R$ monophonic signal is not affected by the presence of the subcarrier.

Next, an $L-R$ signal is generated, as shown in Fig. 1–18. Note when the R signal is passed through a polarity (phase) inverter,

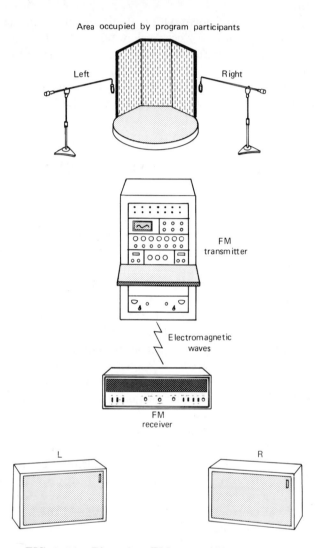

FIG. 1–14.   Plan of an **FM** stereo system.

it is changed into a $-R$ signal. In turn, the $L-R$ signal is produced by adding the L signal to the $-R$ signal. The $L-R$ signal is
amplitude-modulated on the 38-kHz subcarrier and these $L-R$
sidebands are combined with the $L+R$ signal and frequency modulated on the FM carrier. In this manner, the $L-R$ signal has been
encoded into the transmission. As shown in Fig. 1–18(b), the
38-kHz subcarrier itself is suppressed, and a 19-kHz pilot sub-

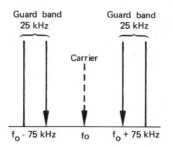

FIG. 1–15.   A standard FM broadcast channel.

carrier is transmitted in its place in order to facilitate trapping out of the (pilot) subcarrier at the stereo receiver. An ordinary mono FM receiver is completely unresponsive to the $L—R$ sidebands and to the 19-kHz pilot subcarrier.

Next, observe how a typical stereo FM receiver operates, as shown in the block diagram of Fig. 1–19. The output from the discriminator branches into an audio amplifier and a bandpass filter. In turn, the filter output contains the $L—R$ sidebands which are combined with a 38-kHz subcarrier from an oscillator. The $L—R$ signal is reconstituted in this way. This $L—R$ signal branches into a mixer and a polarity inverter. Note that the mixer combines the $L+R$ signal with the $L—R$ signal to form a $2L$ signal that drives the L speaker. Then the output from the polarity inverter is a $—L+R$ signal that combines with the $L+R$ signal in a mixer to form a $2R$ signal which drives the R speaker. The 19-kHz pilot-subcarrier trap that picks off the pilot subcarrier for synchronizing the 38-kHz oscillator is not shown in the block diagram.

The foregoing receiver arrangement is called a *matrix decoder design*. In recent years, other types of decoders, called *switching*

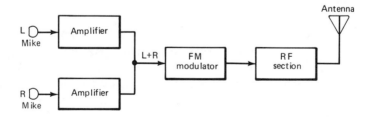

FIG. 1–16.   *L* and *R* microphones generate a conventional monophonic signal.

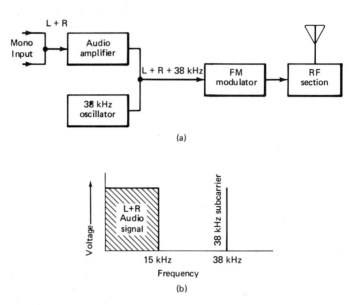

FIG. 1–17.   Subcarrier relation to $L + R$ signal.

bridges and *envelope detectors*, have gained favor. Both are based on the fact that the reconstituted composite stereo has R information along one envelope and L information along the other, as illustrated in Fig. 1–20. Therefore, instead of using a bandpass filter section, the reconstituted stereo signal may be passed through one diode detector to recover the R audio signal and also through another (oppositely polarized) diode detector to recover the L audio signal. This is called the envelope-detection decoder. Again, a semiconductor bridge may be employed, as shown in Fig. 1–21. The diodes conduct alternately and the composite stereo signal is simultaneously reconstituted and rectified to provide L and R audio outputs.

Notice the 67-kHz trap in the configuration of Fig. 1–21. This is called an SCA (subsidiary-carrier-assignment) trap. Some FM stations encode music programs on a 67-kHz subcarrier to provide an additional multiplexed sound channel. This is sometimes called a *storecasting* signal. It is not a hi-fi transmission, in that the 67-kHz subcarrier is close to the 75-kHz channel limit. However, it is satisfactory for "piped-music" systems in commercial establishments. In most cases, SCA programs consist of continuous background music without any station announcements.

Refer to Fig. 1–21 and note R15-C15 and R19-C16. These are RC deemphasis networks. All FM tuners have deemphasis networks because the higher audio frequencies are preemphasized at the FM transmitter in order to obtain a better signal-to-noise ratio in transmission. Figure 1–22 shows the standard preemphasis and deemphasis frequency characteristics. It is necessary to deemphasize the signal after decoding. If the signal were deemphasized before decoding, the decoder would not operate properly because

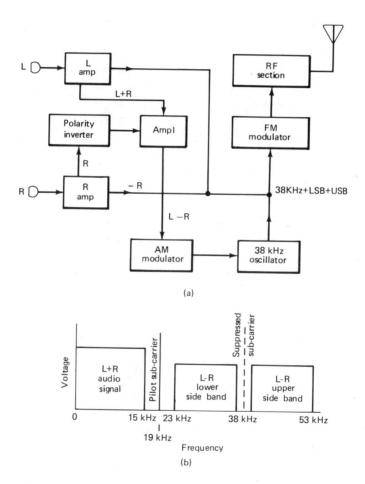

FIG. 1–18. Plan of an FM stereo transmitter. (a) Transmitter block diagram. (b) L + R mono signal and encoded *L — R* signal with 19 kHz pilot subcarrier.

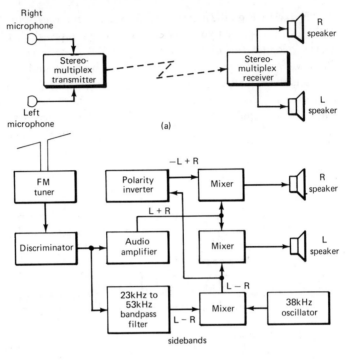

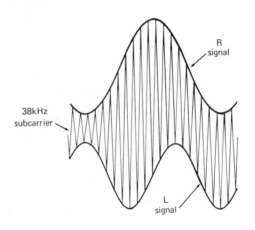

FIG. 1–19. Stereo **FM** reception. (a) System plan. (b) Block diagram of matrix type receiver.

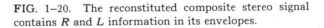

FIG. 1–20. The reconstituted composite stereo signal contains *R* and *L* information in its envelopes.

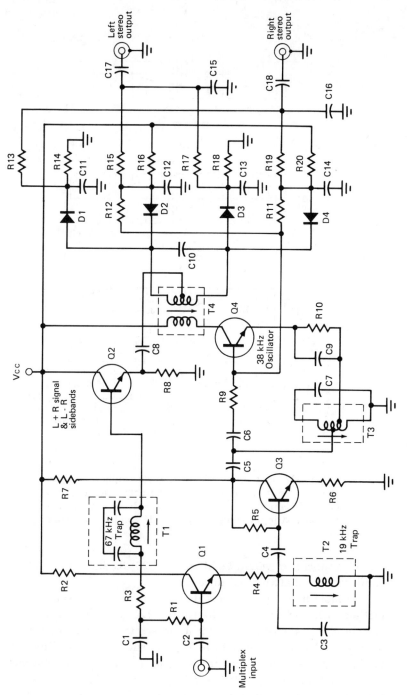

FIG. 1–21. A switching-bridge type of decoder.

21

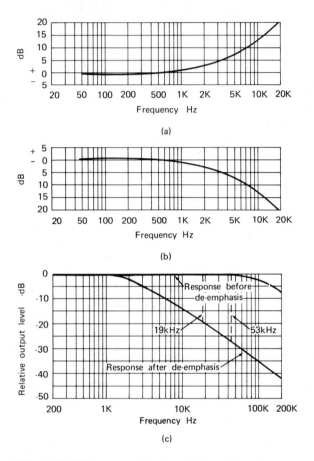

FIG. 1–22. FM pre-emphasis and de-emphasis curves.
(a) Transmitter pre-emphasis characteristic. (b) Receiver de-emphasis characteristic. (c) Receiver frequency
response at input and at output of de-emphasis network.

the pilot subcarrier would be attenuated 20 dB and the *L—R* sidebands would be attenuated up to 30 dB.

Troubleshooting multiplex decoders involves the same general
methods used in servicing FM tuners. DC voltage measurements
are basic. Resistance measurements in semiconductor circuitry are
often aided by employing a hi-lo TVOM, such as the one illustrated
in Fig. 1–23. A hi-lo ohmmeter provides a conventional ohmmeter
test voltage and also a low-test voltage (less than 0.1 volt), so that
resistance measurements can be made without turning semicon-

ductor junctions "on". As an illustration, in Fig. 1–21 the values of $R1$, $R2$, $R3$, $R4$, $R5$, $R6$, $R7$, $R8$, $R9$, and $R10$ can be measured in-circuit with the low-power ohms function of a hi-lo TVOM. Note, however, that a defective transistor or a leaky capacitor could cause a false resistance reading. In other words, an incorrect resistance reading does not necessarily mean that the resistor under test is off-value. A correct resistance reading, of course, clears any suspicion.

1.5   AUDIO AMPLIFIERS

High-fidelity audio amplifiers have a frequency response from 20 Hz to 20 kHz within ± 1dB, and less than 1% harmonic distortion

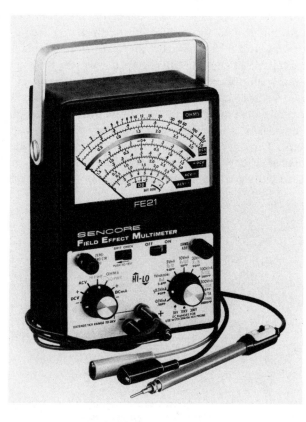

FIG. 1–23.   A hi-lo TVOM. (Courtesy of Sencore)

at maximum rated power output. Power output should be specified in root-mean-square (rms) watts. In the case of a stereo amplifier, the power output should be stated in rms watts per channel. Other power units, such as instantaneous peak power, dynamic power, or IHF power, are less informative and less useful for companies and servicing procedures. A hi-fi amplifier that has less than one percent harmonic distortion will normally have less than one percent inter-modulation distortion at maximum-rated power output. Figure 1–24 shows the relation between harmonic distortion and inter-modulation distortion for a typical amplifier.

It is instructive to note that 1 dB of distortion is difficult or impossible to detect by ear. Similarly, it is practically impossible to see one percent distortion in a sine-wave oscilloscope pattern (Fig. 1–25). Even a five percent distortion is not easy to discern, although a ten percent distortion is quite evident. The only practical method of checking low percentages of harmonic or inter-modulation distortion is with a distortion meter, as detailed shortly.

Although a hi-fi amplifier normally has a frequency response that is flat within ± 1 dB from 20 Hz to 20 kHz, this is the inherent characteristic of the amplifier. In other words, the various inputs of an input amplifier or preamplifier have frequency equalization characteristics that produce attenuation of the higher audio frequencies, as exemplified in Fig. 1–26. Equalization networks have the purpose of providing a flat frequency response for the complete system. This situation is analogous to the deemphasis networks utilized in FM receivers to compensate for the preemphasis of the transmitted signal. Note also in Fig. 1–26 that bass and treble con-

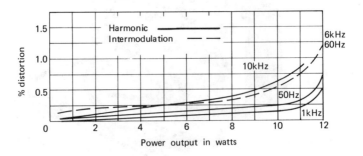

FIG. 1–24. Comparative harmonic and intermodulation distortion curves for a typical amplifier. (Courtesy of General Electric Co.)

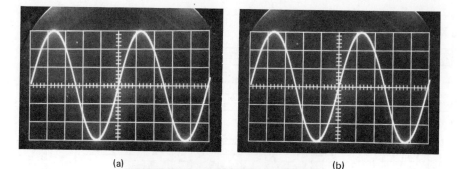

(a)   (b)

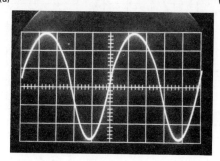

(c)

FIG. 1–25.   Distorted sine waves. (a) 1% harmonic distortion. (b) 5% harmonic distortion. (c) 10% harmonic distortion.

trols are available for adjustment of frequency response to suit personal preferences.

Figure 1–27 shows a configuration for a 15-watt (rms) hi-fi output amplifier. Common trouble symptoms for hi-fi amplifiers include distorted output, weak or no output, noise, hum, or intermittent operation. Distorted output can be caused by leaky capacitors, defective semiconductor diodes, marginal transistor defects, or off-value resistors. In the example shown in Fig. 1–27, the setting of $R2$ should be checked at the outset; an incorrect base-emitter bias voltage will invariably cause distortion. In case there is no output from an amplifier, an oscilloscope serves as a useful signal-tracer. Noise, hum, and intermittent operation can be caused by poor grounding, faulty contacts, or plugs not fully seated into jacks. Noise is occasionally caused by failing transistors. Note that output transistors have a higher failure rate than low-level input transistors.

As noted above, an oscilloscope is not a good indicator of low

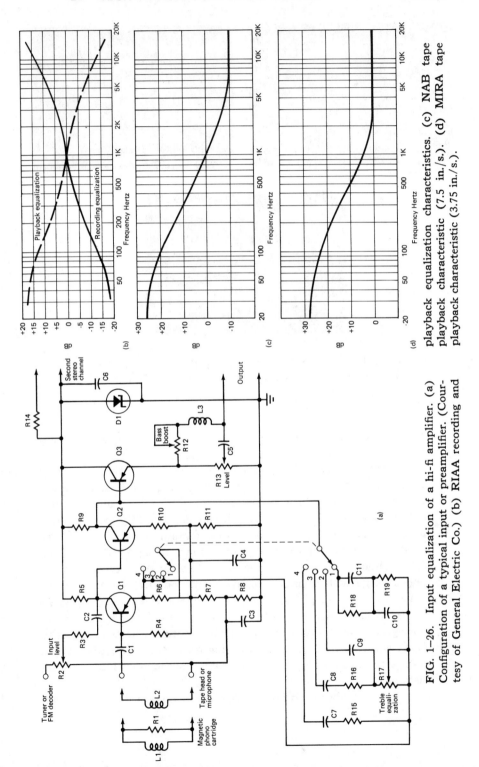

FIG. 1–26. Input equalization of a hi-fi amplifier. (a) Configuration of a typical input or preamplifier. (Courtesy of General Electric Co.) (b) RIAA recording and playback equalization characteristics. (c) NAB tape playback characteristic (7.5 in./s.). (d) MIRA tape playback characteristic (3.75 in./s.).

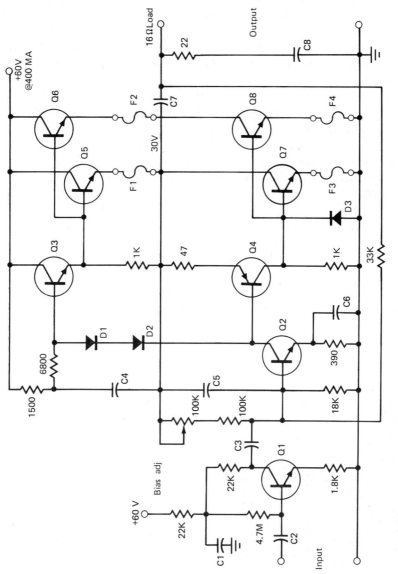

FIG. 1–27. Configuration of a 15 watt hi-fi output amplifier. (Courtesy of General Electric Co.)

percentages of distortion. Accordingly, a harmonic distortion meter must be employed, as shown in Fig. 1–28, to determine whether an amplifier is operating within high-fidelity limits. An intermodulation analyzer may also be utilized, but it is chosen less often because of its comparative complexity. A harmonic distortion meter can be used to good advantage as an output monitor while replacing suspected components in an amplifier. When the defective component is located and replaced, the distortion meter will then indicate a satisfactorily low value, such as 0.5 percent. Note that a harmonic distortion meter measures the effect of amplitude nonlinearity only. It does not indicate whether the frequency response of the amplifier is correct or not. This must be checked with an audio oscillator, as depicted in Fig. 1–29.

Note in Figs. 1–28 and 1–29 that a power resistor R is connected across the amplifier output terminals in place of the speaker. This resistor must be rated for the maximum power output of the amplifier and must have a resistance equal to the rated speaker

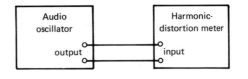

(a)

(b)

FIG. 1–28. Test arrangement for a harmonic distortion check. (a) Audio oscillator waveform check. (b) Amplifier test.

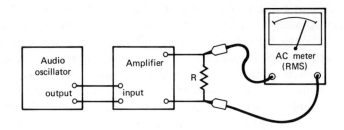

FIG. 1–29.   Check of amplifier frequency response.

impedance. An rms **AC** voltmeter is used to measure the voltage across the load resistor, permitting calculation of the power output. A harmonic-distortion measurement should be made at maximum-rated power output of the amplifier. It is also good practice to check the frequency response of an amplifier at maximum-rated power output.

1.6   HIGH-FIDELITY SPEAKERS

High-fidelity speakers are designed to reproduce frequencies from approximately 30 Hz to 18 kHz. A large cone is required to radiate appreciable power at low frequencies. On the other hand, a large cone cannot reproduce high frequencies efficiently. Therefore, a hi-fi speaker may employ two-way or three-way construction. Two-way speakers have a tweeter unit mounted coaxially with the woofer unit. Three-way speakers (Fig. 1–30) have a mid-range

FIG. 1–30.   A 12 in. 3-way hi-fi speaker. (Courtesy of Radio Shack)

cone and tweeter mounted coaxially with the woofer. This three-way speaker has a frequency response from 24 to 18,500 Hz with a power output of 28 watts. Comparatively high power output is obtained by acoustic suspension of the 12-inch cone. In other words, the cone is mounted in a cloth-roll suspension that permits considerable travel back-and-forth. A ceramic field magnet is also employed to provide an intense magnetic field through the voice coil.

Audiophiles tend to prefer the comparatively expensive *speaker system* design, which utilizes two, three, or more individual speakers in the same cabinet or enclosure. The front panel of a speaker enclosure is called a *baffle board* and is essential in preventing rear radiation of the speaker by canceling the front radiation. If the enclosure has an open back, it is called a *finite baffle*. On the other hand, if the enclosure is air-tight, it is called an *infinite baffle*. Other enclosure designs provide a hole or port in the baffle board to permit some components of the rear radiation to blend with the front radiation. Ports may be ducted or tuned with short pipes or tubes of certain diameters and lengths. The bass-reflex design makes use of a labyrinth (walled-sound channels) behind the speakers to develop the bass-reinforcement (reflex) output. Figure 1–31 depicts a simple, ported enclosure with a basic labyrinth partition. When used with a two-way speaker, it approaches the performance of a speaker system.

Figure 1–32 shows an example of a finite-baffle arrangement with one large speaker and three smaller speakers. This is one type

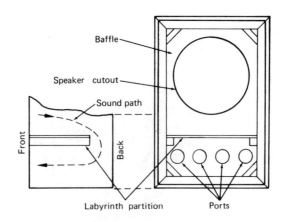

FIG. 1–31. A simple ported enclosure with basic labyrinth partition.

FIG. 1–32.　A finite baffle enclosure with one large and
two small speakers. (Courtesy of Heath Co.)

of speaker system. The individual speakers of a system may be interconnected in various ways. A speaker has a rated value of input
impedance, such as 4, 8, or 16 ohms. Whenever speakers are connected in parallel, the input impedance of the system decreases.
Very low values of input impedance are undesirable from the
practical standpoint. If several speakers with the same impedance
are connected in parallel, the input impedance of the system is
equal to the impedance of one speaker divided by the number of
speakers. Figure 1–33 shows this arrangement. Note that if two
speakers of unequal impedance are connected in parallel, the
speaker with less impedance will draw the most power. Thus, a
four-ohm speaker will draw twice as much power as and sound
louder than an eight-ohm speaker. This is usually an undesirable
situation.

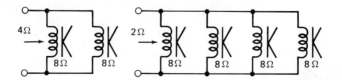

FIG. 1–33. Examples of parallel-connected speaker systems.

Speakers may also be connected in series, as shown in Fig. 1–34. This arrangement increases the input impedance of the system with respect to the impedance of an individual speaker. Note that if one series-connected speaker has a higher input impedance than the other speaker, the high-impedance speaker will draw the most power and sound louder. When neither a parallel arrangement nor a series arrangement provides a good impedance match for the amplifier, a series-parallel arrangement may be employed, as seen in Fig. 1–35. The speaker system should have an input impedance that is reasonably close to the rated output impedance of the amplifier to obtain both maximum power transfer and optimum fidelity of reproduction. In some cases, an amplifier with a low output impedance will be damaged at maximum power output if it is connected to a high-impedance speaker system.

A woofer can handle considerable power, whereas a tweeter is comparatively limited in power capability. Therefore, low-frequency audio currents must be prevented from flowing through the

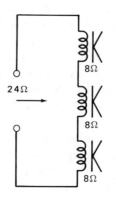

FIG. 1–34.   Series-connected speaker arrangement.

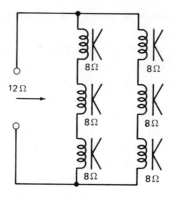

FIG. 1–35. A series-parallel speaker arrangement.

tweeter. This frequency discrimination is accomplished by means of a crossover device or network. The simplest crossover arrangement consists of a capacitor connected in series with the tweeter, as shown in Fig. 1–36. This crossover capacitor does not attenuate high audio frequencies appreciably. On the other hand, we find that attenuation occurs as the frequency is reduced, and at some reduced frequency the reactance of the capacitor will be equal to the impedance of the tweeter. This is the cross-over frequency. In the example of Fig. 1–36, the crossover frequency is about 5 kHz. An 8-μF capacitor would provide a crossover frequency of approximately 2.5. kHz. At the crossover frequency, half the source power is applied to the tweeter. Practically no power is applied to the tweeter at low audio frequencies. The value of the crossover

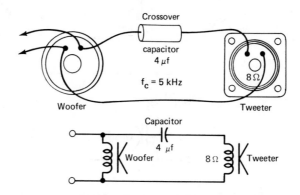

FIG. 1–36. Example of a crossover capacitor arrangement.

capacitor is chosen to provide a system frequency response that is as uniform as possible.

In case the woofer is capable of reproducing frequencies above the crossover frequency, there is a tendency to overemphasize the midrange with a simple crossover capacitor. Therefore, an LC crossover network is often used, as exemplified in Fig. 1–37, to obtain better uniformity of frequency response. In practice, the value of the coil is adjusted as required for optimum operation. An LC crossover network has another advantage in that the system input impedance remains fairly constant over the entire audiofrequency range. To achieve tonal balance, it is generally desirable to connect a variable resistance (potentiometer) in series with the tweeter. A 50-ohm, 2-watt potentiometer is often used with a cone-type tweeter. It is adjusted to give a treble-loudness level equal to the bass-loudness level. This is essentially a subjective judgement and the adjustment is made on the basis of listening to various musical passages.

Speakers operating in the same enclosure must be driven in phase with each other. Otherwise, the output from one speaker will tend to cancel the output from the other speaker. When speakers operate in phase, their sound outputs are additive. Figure 1–38 shows how speakers are connected to operate in phase for series and parallel arrangements. If red dots are not used to identify corresponding terminals on speakers, a test can be made with a dry cell to determine if

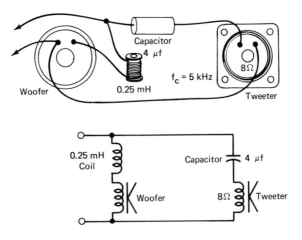

FIG. 1–37.   An LC crossover network.

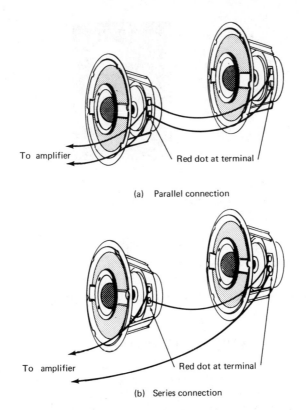

(a)   Parallel connection

(b)   Series connection

FIG. 1–38.   Connection of speakers for in-phase operation.

the speakers are operating in phase. When the dry cell is connected to the speaker in one polarity, the cone will move forward slightly. If the polarity is reversed, the cone will move backward slightly. The speakers are connected in phase when the dry-cell voltage makes both cones move in the same direction.

When more than one speaker is mounted in an enclosure, the woofer should be placed at the bottom and the tweeter at the top, as exemplified in Fig. 1–39. If a port is employed, it should be located nearer the woofer than the tweeter. To avoid cabinet resonance, the inner surface of the enclosure should be considerably padded with a sound-absorbing material such as fiberglass. If speaker enclosures are operated in a highly-reflective room, such as a rumpus room, it is often advantageous to utilize closed-back

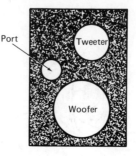

FIG. 1–39. Suitable baffle-board layout for woofer, tweeter, and port.

cabinets placed as depicted in Fig. 1–40. On the other hand, in a room that is highly sound-absorbent, the cabinets generally may be placed in opposite corners of the room, as shown in Fig. 1–41. Either open- or closed-back cabinets are suitable.

1.7 QUADRIPHONIC SYSTEMS

A quadriphonic sound system utilizes four speakers, as shown in Fig. 1–42. Audio buffs generally energize the speakers from four-channel quadriphonic tapes. This is called discrete or true four-channel operation. It is similar to stereophonic reproduction, with the addition of two speakers at the rear of the listening area. These rear speakers have the function of reproducing the reflected sound that is heard by the audience in a concert hall. Quadriphonic reproduction gives the reflected sound sources their approximately

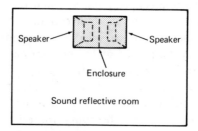

FIG. 1–40. Stereo speakers placed in a highly reflective room.

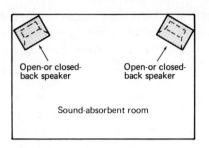

FIG. 1–41.   Stereo speakers placed in a sound-absorbent room.

correct locations with respect to the listener. In addition to four-channel quadriphonic tapes, four-channel records are also available. These are essentially stereo recordings with encoded information in each of the stereo channels. A decoder is utilized with the quadriphonic amplifier system to "sort out" the four audio signals.

Although less favored than discrete quadriphonic sound, syn-

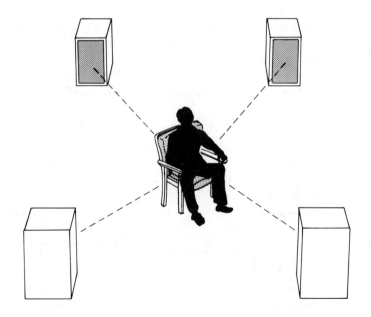

FIG. 1–42.   Placement of quadriphonic speakers.

thesized four-channel sound is also utilized. A synthesizer operates in combination with two-channel records, tapes, or FM stereo to derive four-channel information from a two-channel source. Although complete reconstitution is impossible, partial reconstitution is obtained on the basis of phase difference between a signal reproduced by the *L* channel and the *R* channel. RC networks are used in a synthesizer to develop the reconstituted audio channels. Fig. 1–43 shows how a four-channel adapter (synthesizer) is connected between a source of stereo signals and the four speakers in a quadriphonic system. Audio buffs occasionally find need for resistive pads or attenuators in stereo or quadriphonic systems. Values for basic *T* and *H* pads are given in Fig. 1–44. For impedances other than 200, 500, or 600 ohms, the values are proportional and may be found by interpolation.

It is generally agreed that the best reproduction of 4-channel sound is provided by four separate channels from the recording studio to the four speakers in the listening area. However, it is not always possible to provide four separate channels. As an illustration, FM stations cannot broadcast discrete 4-channel sound at this writing, owing to technical limitations imposed by FCC regulations. Audio tape is the most widely used source of discrete 4-channel sound. Recording and playback arrangements are shown in Fig. 1–45. Since the tape employs four tracks, its playing time is one-half that of a stereo tape. Note that tapes are more costly than discs with equal playing time.

Disc records that provide four discrete channels have been developed. One technique utilizes a frequency-modulated supersonic

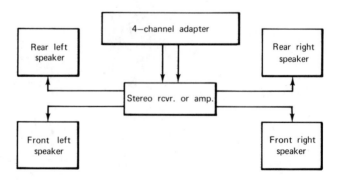

FIG. 1–43.   A four-channel synthesizer (adapter) is connected between a source of stereo signals and four speakers.

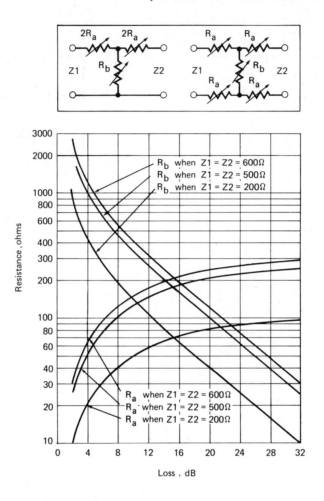

FIG. 1–44.   Resistance values for basic *T* and *H* attenuators and pads.

carrier added to each channel of a stereo recording. The conventional *L* and *R* stereo channels provide the sum of the front and
rear sounds; they reproduce the conventional stereo sound when
played on ordinary stereo record players. These supersonic carriers
are modulated by the difference signals. In turn, after the carriers
are demodulated and mixed with the right and left channel signals,
four separate discrete channels of sound are obtained. It is impractical at this time for **FM** stations to broadcast this quadriphonic
signal because it is very difficult to maintain adequate separation
of the channels.

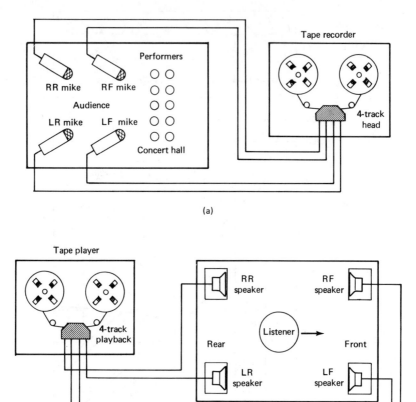

FIG. 1–45.   Recording and playback arrangements for discrete 4-channel sound. (a) Recording. (b) Reproduction.

A block diagram of the circuit sections used in playing back 4-channel discrete quad records is shown in Fig. 1–46. The 30-kHz supersonic carrier is frequency-modulated by the stereo-difference signal in the recording. In the playback process, the audio signals from the two conventional stereo channels are mixed with the signals from the demodulated carriers. This mixing action reconstitutes the original four discrete signals which are then amplified and fed to four speakers. Another approach employs encoders during recording

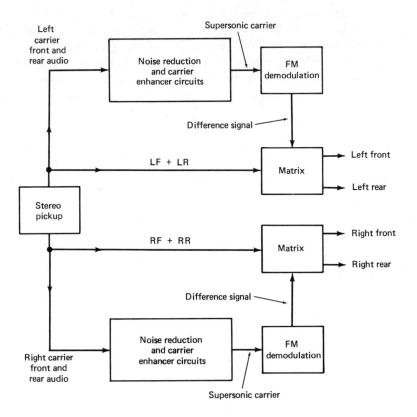

FIG. 1–46. Block diagram of playback system for 4-channel discrete-stereo records.

and complementary decoders during playback, as shown in Fig. 1–47. Various types of encoder/decoder circuits are utilized. A widely used arrangement is depicted in Fig. 1–48. This method encodes audio signals by means of 180° phase shifts. It has a disadvantage, however, of cancellation of some audio frequencies in localized listening areas.

    To avoid the foregoing 180° phase-shift system, the QS quadriphonic-stereophonic system has been developed. It features 90° phase shifts of encoded signals, as shown in Fig. 1–49. This QS system, in turn, has the disadvantage of incomplete separation. To obtain improved separation, gain-riding circuitry is utilized in the more elaborate versions of the QS system. These gain-riding circuits automatically exaggerate the separation between channels whenever a separation is sensed. QS systems are compatible with conventional stereo equipment. However, the compatibility is not quite complete, and when a stereo record is played through a 4-channel QS system, a synthetic diffused-stereo reproduction results. Some listeners approve of diffused-stereo reproduction, and others assert that the sound is unnatural.

Note that the final matrices are not shown in Figs. 1–48 and 1–49. In other words, the *LF, LR, RR,* and *RF* outputs are mixed in matrices to recover the Left, Right, Front and Rear audio information. Thus, $LF + LR = 2L; \; RR + RF = 2R; \; LR + RR = 2R$

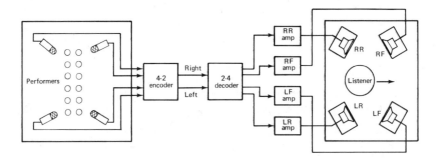

(a)

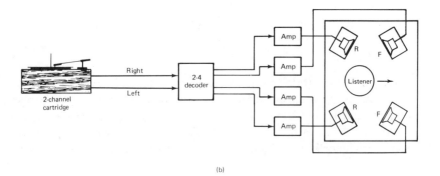

(b)

FIG. 1–47. Encoder-decoder quadriphonic arrangement. (a) For 4-2-4 systems in stereo-FM or stereo tapes. (b) A complementary decoder provides four channels during playback.

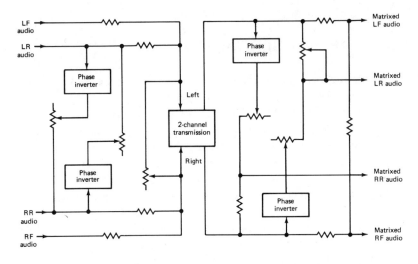

FIG. 1–48. A 180° phase-shift encoder-decoder arrangement.

(Rear); $LF + RF = 2F$ (Front). It is evident that many design variations are possible in encoded quadriphonic systems, all of which have certain disadvantages at this time. Possibly, some new technical approach will be discovered which could minimize or eliminate the problems that now burden quadriphonic sound systems.

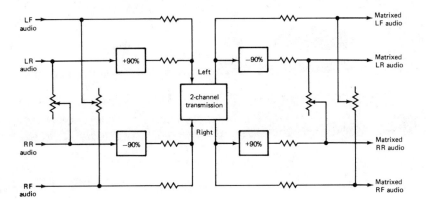

FIG. 1–49. A 90° phase-shift encoder-decoder arrangement.

1.8   TAPE RECORDERS

A tape recorder has its own amplifier, microphones, and speaker(s); a stereo recorder, such as illustrated in Fig. 1–50, provides two recording and playback channels. This is a cassette-type recorder, which provides compactness and operating convenience in comparison to open-reel type recorders. Note that a cassette contains two tape reels which are enclosed in a plastic case. A cartridge has a single reel, designed to unwind from the center as it rewinds at the circumference. A cassette can be flipped over and the tape thereby rewound from one reel to the other. On the other hand, a cartridge cannot be flipped over since the tape must always move in the same direction. Cassette tapes generally run at $1\frac{7}{8}$ in./sec and cartridge tapes usually run at $3\frac{3}{4}$ in./sec.

Standard track placements on stereo and quadriphonic tapes are shown in Fig. 1–51. The configuration of a typical tape-recorder amplifier is seen in Fig. 1–52. It consists of a four-stage audio amplifier and a 32-kHz bias oscillator. This AC bias voltage is employed for recording tapes. Note that $M1$ is a combination

FIG. 1–50.   A stereo cassette tape recorder. (Courtesy of Radio Shack)

record-playback head and $M2$ is an erase head. In this example, the erase head utilizes DC voltage. In other designs, the erase head is energized by a comparatively high-level AC voltage. Note also that some tape recorders employ separate record and playback heads. An AC bias voltage is always applied to the record head to obtain maximum amplitude linearity (minimum harmonic distortion). Standard recording tape utilizes a coating of red iron oxide. Low-noise tape employs chromium dioxide. Since a higher AC bias voltage is required for chromium-dioxide tape, a bias-level switch may be provided.

A record head has a somewhat wider gap than a playback head. Thus, a combination head has a compromise gap width. In practice, gap widths range from 0.006 to 0.010 inch. An erase head has a large gap width. Heads are subject to wear due to friction of the tape moving across the head. When a gap is sufficiently worn

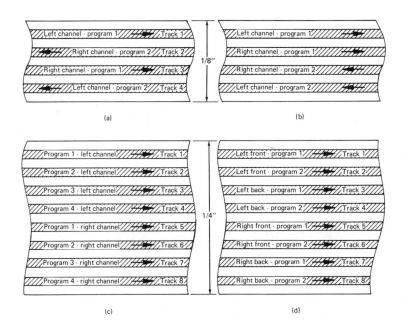

FIG. 1–51. Standard track placements on stereo and quadriphonic tapes. (a) Reel-to-reel (open-reel) tape, stereo recording. (b) Cassette tape, stereo recording. (c) Eight-track cartridge tape, stereo recording. (d) Eight-track cartridge tape, quadriphonic recording.

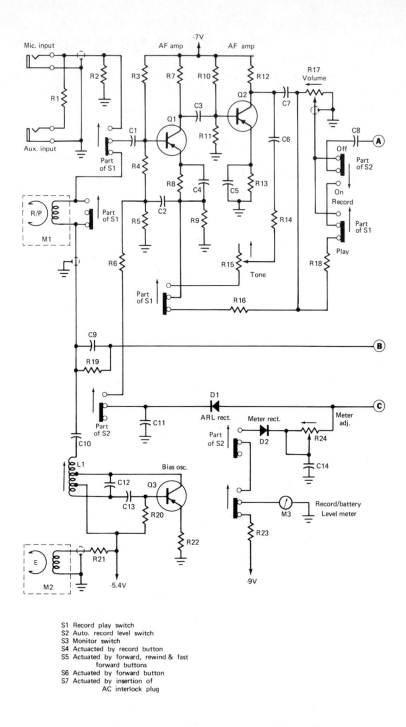

FIG. 1–52. An amplifier and bias oscillator configuration for a tape recorder.

S1 Record play switch
S2 Auto. record level switch
S3 Monitor switch
S4 Actuacted by record button
S5 Actuated by forward, rewind & fast
 forward buttons
S6 Actuated by forward button
S7 Actuated by insertion of
 AC interlock plug

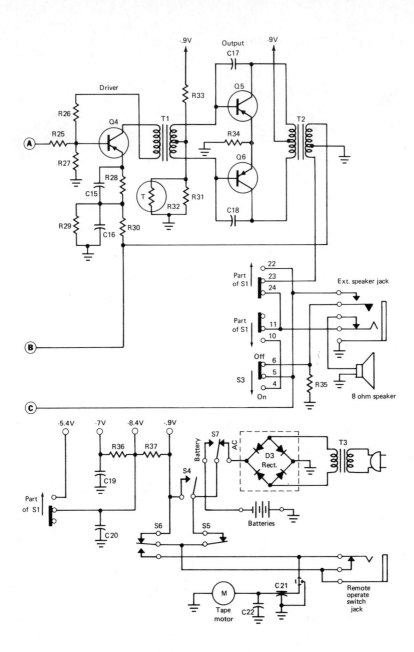

FIG. 1–53. Typical tape heads. (Courtesy of Nortronics Co., Inc.)

that it increases in width, the head must be replaced. Figure 1–53 shows the appearance of various types of heads. It is also essential to keep tape heads clean so that the tape comes into close contact with the gap. After an appreciable length of time, a head is likely to become magnetized gradually. Since this magnetization reduces the quality of reproduction and tends to damage the information on the tape, heads should be routinely demagnetized by an appropriate device that applies a substantial AC field to the head.

When a head is replaced in a recorder, it is essential to mount the replacement at the precise correct height and with correct azimuth. In other words, the gap should be perpendicular to the tape for best results. If the head leans slightly to the left or right, reproduction will be impaired. Test tapes are commonly used to check the azimuth adjustment. An output meter will indicate when the adjustment is correct by a peak maximum indication. Since there are minor peaks that occur on either side of the major peak, the technician should make certain that a peak maximum indication has been obtained. The output from a playback head is comparatively low, and may normally be less than 1 mV. For this reason, a high-gain amplifier is required. The same amplifier is used both for recording and for playback.

Troubleshooting of tape-recorder amplifiers involves the same approaches and tests previously described for general high-fidelity amplifiers. When a transistor is suspected of being defective, it is informative to make in-circuit turn-off and turn-on tests. In other words, transistors are usually soldered into printed circuits, and it is time consuming and laborious to remove a transistor for test. Therefore, in-circuit tests are very helpful. Figure 1–54 shows how turn-off and turn-on tests are made. In a turn-off test, the base and emitter of a transistor are temporarily connected together. This causes the base and emitter voltages to be the same and a normally operating transistor *turns off* or stops drawing collector current.

Therefore, the collector voltage jumps up to the supply-voltage value during the turn-off test. Failure to respond in this manner indicates that the transistor is defective.

In a turn-on test, some of the collector voltage is bled into the base circuit of the transistor with a test resistor. This has the effect of increasing the forward bias on the transistor, with the result that an increased collector current is normally drawn. Therefore,

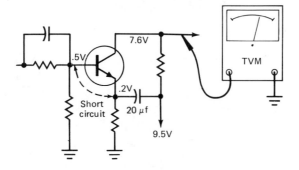

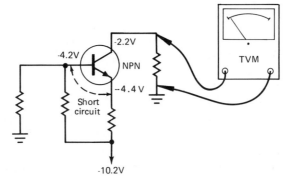

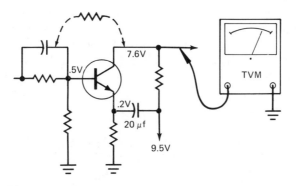

FIG. 1–54. Transistor turn-off and turn-on tests.

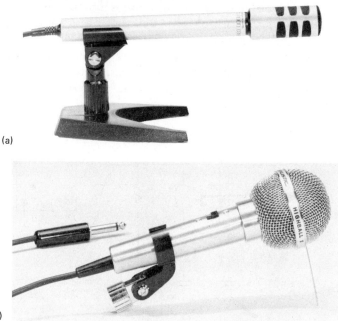

FIG. 1–55. Typical recorder microphones. (a) Dynamic microphone. (b) Capacitor microphone. (Courtesy of Radio Shack)

the collector voltage decreases during the turn-on test. Failure to respond in this manner indicates that the transistor is defective. Some circuits have configurations where turn-off and turn-on tests are impractical. However, the majority of circuits lend themselves to these tests. As an illustration, $Q1$, $Q2$, and $Q3$ in Fig. 1–52 are easily checked in this manner. $Q4$ and $Q5$ can also be checked, but only a turn-off test should be made in this situation because it is possible to damage power transistors in a turn-on test. Since this is a push-pull circuit, it is advisable to short-circuit the base and emitter of both $Q4$ and $Q5$ at the same time. With both transistors normally cut off, the collector voltages will rise to the supply-voltage value (an increase of 0.1 volt).

1.9   MICROPHONES FOR TAPE RECORDERS

Most tape recorders are used with dynamic or with capacitor microphones, as illustrated in Fig. 1–55. In these examples, the dynamic

microphone has a rated frequency response from 100 to 10,000 Hz
and an output impedance of 10,000 ohms. The capacitor microphone has a rated frequency response from 30 to 15,000 Hz and
an output impedance of 600 ohms. Some dynamic microphones
provide a choice of output impedances, such as 600 and 20,000
ohms. The output voltage of a recorder microphone is comparable
with the output voltage of a tape head. Figure 1–56 depicts the
construction of a dynamic microphone. It consists of a diaphragm
with a moving coil suspended in an air gap traversed by a strong
magnetic field. Since the output impedance of the moving coil is
very low, a step-up microphone transformer is used to increase the
output impedance. A protective screen woven of fine wire mesh
improves the acoustic response by minimizing breath sounds, *pops*,
and booming.

The construction of a capacitor microphone is shown in **Fig.**
1–57. It comprises an air-dielectric capacitor. The front plate is
very thin and serves as a diaphragm. Spacing between the plates
is typically 0.001 inch. The diaphragm is stretched by its mounting arrangement to obtain a mechanical resonant frequency in the
range of 5 to 10 kHz. In this manner, maximum uniformity of frequency response is obtained. An electrostatic field is established
between the plates by a battery potential, typically of 180 volts. A
series resistor of approximately 10 megohms is utilized in series
with the battery. In turn, when the diaphragm vibrates, an **AC**
voltage is developed between the plates. Because the internal impedance of a capacitor microphone is very high, an **FET** preamp
is often built into the microphone housing, as shown in **Fig. 1–52.**

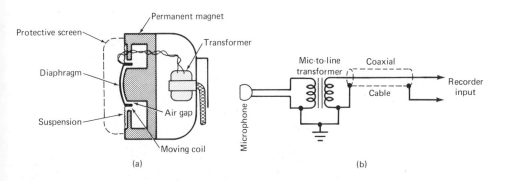

FIG. 1–56. Construction of a typical dynamic microphone. (a) Plan of transducer. (b) Transformer arrangement.

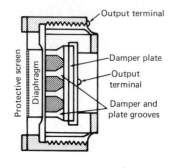

(a)

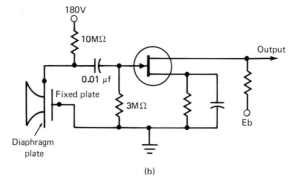

(b)

FIG. 1–57. Construction of a typical capacitor micro-phone. (a) Plan of transducer. (b) Preamp arrangement.

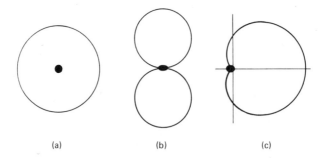

(a)   (b)   (c)

FIG. 1–58. Basic directional patterns for microphones. (a) Omnidirectional pattern. (b) Bidirectional pattern. (c) Cardioid pattern.

This preamp serves to provide a conventional output impedance and also to increase the audio output level from the microphone.

Various microphones are designed to provide particular directional patterns. Of these, the omnidirectional, bidirectional, and cardioid patterns are basic, as shown in Fig. 1–58. Omnidirectional microphones are useful chiefly in monophonic recording, where there is no problem in extraneous noise fields. Since an omnidirectional microphone responds equally from all directions, it can be used without difficulty by inexperienced personnel. Bidirectional microphones are useful in stereo recording since they have an axis of maximum response and an axis of minimum response at right angles. Better separation of sound sources can be obtained under some conditions in this way. A cardioid microphone is useful for both stereophonic and quadraphonic recording. It has maximum forward response and minimum rearward response. Note that a cardioid microphone has a sound pattern that tends to change with frequency. In other words, at low audio frequencies, the cardioid pattern gradually opens out and approaches an omnidirectional pattern.

1.10   DOLBY NOISE-REDUCTION SYSTEM

The Dolby noise-reduction system exploits dynamic signal compression and expansion during recording and playback processes to reduce the noise level inherent in these processes. It is operative in the high audio-frequency region, which has a considerably higher noise potential than the low-frequency region. Volume expansion has been employed for many years in various audio systems. Volume compression has the inverse effect. Both processes tend to introduce harmonic and intermodulation distortion; however, with proper design, these undesirable side effects of expansion and compression can be kept to a low value. The simplest volume-expansion arrangement incorporates reverse automatic gain-control action with an audio amplifier. The response time of the agc network is quite slow compared to an AGC network in a radio receiver. When the amplitude of the program material increases, the gain of the amplifier increases. Or when the program amplitude decreases, the amplifier gain decreases. This volume-expansion process helps to mask noise, as noted previously.

Volume compression also incorporates agc action arranged to decrease the amplifier gain as the amplitude of the program material increases. Or when the program amplitude decreases, the amplifier gain increases. If volume expansion is utilized during recording and an inversely proportional volume compression is employed during reproduction, the amplitude input-output relationship is linear. When these processes are employed in an optimum design, the signal-to-noise ratio of the system is significantly improved. Inevitably, there is a trade-off involved, in that the percentage distortion of the system becomes increased at least slightly, as noted above.

Noise reduction can also be obtained by reducing the high-frequency response of the system. For example, an amplifier may be designed to attenuate frequencies starting at 6 kHz with practically zero output at 10 kHz. Although this simple method effectively reduces the noise level, it is not a high-fidelity approach. Another method of noise reduction employs high-frequency preemphasis in recording with high-frequency deemphasis in reproduction. This approach has been previously described under FM transmission and reception. Still another method of noise reduction, called the Olson noise suppressor, employs an amplifier that has no response until the signal level exceeds a certain threshold value, such as 0.5 volt. This approach is based on the fact that low-level program material has the poorest signal-to-noise ratio. A trade-off is involved, of course, inasmuch as audio signals below the threshold value are not reproduced, and some intermodulation distortion occurs. If four channels are employed in the Olson system, no frequency discrimination is involved.

The Scott dynamic noise suppressor exploits bandwidth control and is based on the fact that the noise output of a system is proportional to its bandwidth. Separate high-frequency and low-frequency tone controls are utilized and are automatically controlled by the audio signal. The design approach is to restrict the bandwidth of the amplifier during the time that it is not actually required by the audio information. In other words, if there is little or no high-frequency information present, the automatic treble control reduces the high-frequency response of the amplifier. On the other hand, if there is little or no low-frequency information present, the automatic bass control reduces the low-frequency response of the amplifier. In case there is mid-range information only, both the high-frequency and the low-frequency response of the amplifier are reduced.

FIG. 1–59. Typical automatic turntable. (Courtesy of Radio Shack)

FIG. 1–60. Stroboscopic check of turntable speed. (Courtesy of General Industries Co.)

1.11 TURNTABLES AND PHONO CARTRIDGES

A typical high-fidelity turntable is illustrated in Fig. 1–59. Basic speeds for turntables are 78, 45, 33⅓, and 16⅔ rpm. However, 78-rpm discs are obsolete. Music is not reproduced satisfactorily by 16⅔-rpm discs. High-fidelity recordings are provided by 45-rpm, 7-inch, and 33⅓ 12-inch discs. Turntable speed is conveniently checked with a strobe pattern, as shown in Fig. 1–60. A strobe check is also useful for detecting wow (any variation in turntable speed). Other basic considerations are tracking pressure and anti-skating characteristics. Tracking pressure denotes the amount of force or weight that the stylus places on the record groove. It is measured in grams with a stylus pressure gage such as illustrated in Fig. 1–61. A maximum pressure of 5 grams is employed in high-fidelity reproduction. Some cartridges and styli will track satisfactorily with approximately 1 gram of pressure, provided a high quality tone arm is used.

Skating denotes any tendency of a tone arm to move when the turntable is not exactly level or when a warped disc is being played. A precisely balanced tone arm does not have any skating tendency; this condition is provided by an anti-skate balancing device. A high quality turntable is comparatively heavy; it must be strongly constructed. Inferior turntables are subject to wow, rumble, and sometimes hum when certain types of cartridges are used. Rumble results from motor and turntable vibrations that are mechanically channeled through the mounting bars and the tone arm to the cartridge. Rumble sounds like furniture being moved around upstairs. Wow is reproduced as a low-frequency modulation of the audio information. Some low-level cartridges, particularly the moving-coil type, cannot be used satisfactorily

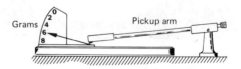

FIG. 1–61.   A stylus pressure gage.

with steel turntables. Electromagnetic pickups are particularly prone to hum pickup with inferior turntables.

An automatic turntable typically plays a stack of 12-inch or 7-inch discs sequentially on one side and then turns off the power switch. A more elaborate automatic turntable plays a stack of intermixed 12-inch and 7-inch discs sequentially on one side. It is good practice to adjust the vertical angle that the stylus makes with the discs so that the departure from the normal angle is the same, although opposite, for the first disc and the last disc of a stack. Normally, this angle will be correct unless a replacement cartridge is improperly mounted in the tone arm. Serious audiophiles generally prefer manually operated turntables because no compromises are involved and any high-fidelity cartridge can be utilized. Hum problems and rumble are seldom encountered with manually operated turntables.

Although various types of cartridges have been manufactured, the magnetic type is in widest use. A hi-fi type of diamond-stylus magnetic stereo cartridge is typically rated for a frequency response of 10 to 25,000 Hz. A stereo cartridge contains two transducers for L and R channel reproduction. The diamond stylus generally has an elliptical contour with tip dimensions of 0.0002 $\times$ 0.0007 in. This type of stereo cartridge provides 25-dB channel separation at 1 kHz and satisfactorily tracks with a stylus pressure of $3/4$ to $1\frac{1}{2}$ grams. As a rule of thumb, high-fidelity reproduction is obtained from a cartridge at a comparatively low output level. In turn, sufficient amplification must be provided to achieve the desired audio-power output level from the system. By way of comparison, a low-level cartridge produces an output of less than 20 mV, whereas a high-level cartridge provides more than 100 mV output. A variable-reluctance cartridge has an output of approximately 10 mV. Output impedances of magnetic cartridges range from 7,000 to 100,000 ohms.

So-called permanent styli have diamond, sapphire, or ruby tips. Tungsten carbide is also used, in addition to osmium and various metal alloys. The chief disadvantage of a diamond stylus is its high cost. However, diamond provides the longest life. Sapphire tips wear out comparatively rapidly, tungsten carbide has a reasonably long life, but osmium-tipped styli are comparatively short-lived—approximately 25 percent that of sapphire. When the stylus moves laterally in a groove, it encounters a narrowing of the channel. This produces a *pinch effect* with incidental distortion. The accepted method of minimizing pinch distortion is to use an elliptical stylus, as noted previously. The pinch effect can be disregarded in playing mono discs, but is significant with stereo discs because no vertical motion of the stylus is tolerable in stereo reproduction. Stylus wear is checked with a small microscope, as illustrated in Fig. 1–62. A cartridge should be replaced when the stylus shows visible evidence of a *flat*. The chief disadvantages of the elliptical stylus are its comparatively high cost and its shorter useful life as compared to an equivalent conical stylus.

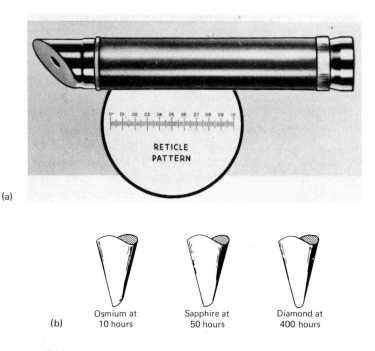

(a)

(b)

| Osmium at 10 hours | Sapphire at 50 hours | Diamond at 400 hours |

FIG. 1–62.   Stylus inspection microscope. (a) Appearance of microscope. (b) Comparative needle wear. (Courtesy of Edmund Scientific Co.)

2

INTER-COMMUNICATION UNITS AND HIGH-FIDELITY WIRING SYSTEMS

2.1   GENERAL CONSIDERATIONS

Residences, offices, shops, and factories may be wired for inter-communication and/or high-fidelity reproduction of FM radio (including SCA programs), tapes, and records. In other words, many of these sound systems combine intercommunication and radio program facilities. If high-quality installations are utilized, high-fidelity reproduction can be provided. Note that intercommunication systems may be of the wired or the wireless type. Wireless intercommunication exploits a form of carrier-current transmission that is conducted by the lighting and power circuits in a building. In most locations, carrier-current sound reproduction is comparatively noisy and may be unacceptable for entertainment purposes. Therefore, most installations employ special cable runs for intercom and music reproduction. The National Electrical Code classifies this type of wiring system under the remote-control, low-energy power, low-voltage power, and signal category. Although the code requirements for this type of system are greatly relaxed in comparison with those for light and power wiring, nevertheless there are certain good practices which should be observed. Otherwise, noise, interference, and frequency distortion may defeat the purpose of the installer.

The simplest intercommunication arrangement, and one which is in fairly wide use on construction projects and in remote areas consists basically of a pair of telephone receivers with a two-conductor line, as depicted in Fig. 2–1. The telephone receiver at each end of the line serves alternately as a transmitter and as a receiver. As seen in Fig. 2–1(b), the receiver consists essentially of a permanent magnet with a coil wound around its end, which is placed close to a thin, iron diaphragm. When sound waves strike the diagragm, it is set into vibration and audio currents are generated in the coil. Conversely, when audio currents are applied to the coil, the receiver

(a)

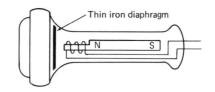

(b)

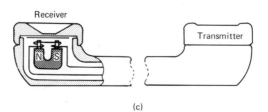

(c)

Fig. 2–1. The simplest intercommunication arrangement. (a) Receiver connections to line. (b) Construction of basic receiver. (c) Conventional form of receiver.

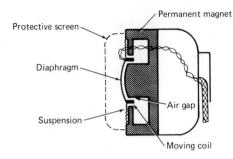

Fig. 2–2. Moving-coil type of transducer.

reproduces the original sound waves from its diaphragm. The conventional form of receiver is depicted in Fig. 2–1(c). It consists of a horseshoe-type permanent magnet with a pair of coils wound around the pole pieces of the magnet. The pole pieces are placed near a thin iron diaphragm, as in the basic receiver.

When a receiver and transmitter are combined in a single housing, as exemplified in Fig. 2–1(c), the arrangement is called a handset. The transmitter in a handset may be a duplication of the receiver construction or it may have a different and specialized construction. In the case of a sound-powered system, the transmitter usually has the same construction as the receiver. Note that a sound-powered intercommunication set does not use batteries or any other source of electricity. In other words, there is no voltage on the line unless sound waves strike the diaphragm of the transmitter. The transmitter converts sound energy into electrical energy; hence, the name "sound-powered equipment". There has been a marked trend toward using the moving-coil type of transducer in sound-powered intercom sets, as shown in Fig. 2–2. This type of transducer was described in the first chapter.

Some economy-type intercom sets provide more signal power than a sound-powered arrangement by using carbon microphones (transmitters), as depicted in Fig. 2–3. A carbon microphone contains carbon granules in a chamber between a pair of carbon discs. One of the carbon discs vibrates when sound waves strike the diaphragm. In turn, the resistance of the granules changes and the amount of battery current changes through the microphone. A carbon microphone provides amplification of the impressed audio energy both because of the battery current and because of the characteristics of the carbon granules. This increased output energy permits the use of longer lines when a carbon microphone is used

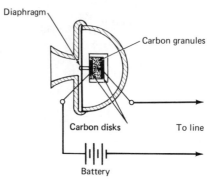

Fig. 2–3. Carbon microphone (transmitter) construction.

instead of a sound-powered transducer. Note in passing that elaborate intercom systems always use electronic amplifiers to increase the output energy from a transducer.

Figure 2–4 illustrates a commercial form of sound-powered intercom unit. For calling purposes, it may be supplemented by a magneto-type ringer. In other words, a small magneto may be utilized to generate a comparatively strong calling signal. The magneto consists of a permanent magnet and a coil which can be vibrated rapidly over the pole of the magnet by means of a lever or of a push-button. In turn, the receiver at the far end of the line emits a loud calling signal. Since the magneto is operated by manual

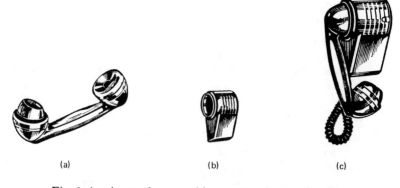

(a) (b) (c)

Fig. 2–4. A sound-powered intercom unit. (a) Handset. (b) Handset holder. (c) Handset holder with magneto ringer. (Courtesy of Wheeler)

energy, no battery or other source of electricity is required. This is a distinct advantage in remote areas where battery replacement can be a serious problem. It is instructive to note that the first telephone invented by Alexander Graham Bell was a sound-powered design.

Some intercom sets supplement loudspeaker facilities with earphone reproduction. An earphone is useful in noisy locations and also in quiet locations when the incoming signal is weak. Another advantage of an earphone is the privacy that it provides. Figure 2–5 shows the basic construction of an earphone. Its housing is sometimes called a *watchcase receiver* because of its size. Some earphones are extremely small and may be worn in the ear. With reference to Fig. 2–5, *D* is the thin, iron diaphragm, *P* is a permanent magnet, *N* and *S* are the pole pieces of the magnet, and *CC* are the coils wound over the ends of the poles. The coils are wound with many turns of rather thin wire and have a winding resistance in the range of 50 to 150 ohms.

By way of comparison, note that a carbon microphone such as depicted in Fig. 2–3 has an internal resistance of 50 to 75 ohms, while a moving-coil microphone such as shown in Fig. 2–2 has an internal impedance in the range of 4 to 16 ohms. Since maximum power is transferred when impedances are matched, we may find matching transformers included in intercom sets.

2.2   INTERCOMMUNICATION EQUIPMENT

Intercommunication units are classified into master and remote types. A master station can call any one of the remote stations. On the other hand, a remote station can call the master station only. An intercommunication wiring system may employ master stations

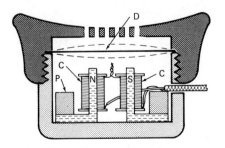

Fig. 2–5.   Construction of an earphone.

throughout. In this case, any station can call any other station. Figure 2–6 illustrates a simple intercom set comprising one master and two remote stations. Switching facilities are provided in this example for as many as four remote stations. Note that a remote station may be installed indoors or outdoors. Thus, a remote station might be installed in the laundry or on the front porch. This type of intercom set has remote units with "beep" buttons that permit a caller to signal the master station even when the system is turned "off".

Figure 2–7 depicts plans for typical intercom wiring systems. An extremely large variety of wiring systems can be devised to meet individual installation requirements. Another type of intercom set has a resemblance to a telephone handset, as exemplified in Fig. 2–8. In this example, facilities are provided for separate conversations between 16 stations simultaneously. Conference calls can also be utilized with up to 16 individuals participating. No central exchange is employed (100 percent trunkage is provided). Figure 2–8 also shows the plan of an eight-station installation for this type of intercom. Here a separate power supply is used which powers all stations. Multiconductor cable is required for the wiring system, as detailed subsequently.

The configuration for a typical five-station intercom is shown in Fig. 2–9. Five master stations may be employed in the system or a minimum of one master station and four substations. Calls can be originated only at master stations. A seven-conductor cable is

Fig. 2–6. An intercom set comprising one master and two remote stations. (Courtesy of Heath Co.)

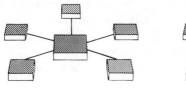
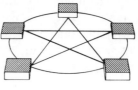

(a)

(b)

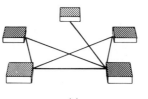

(c)

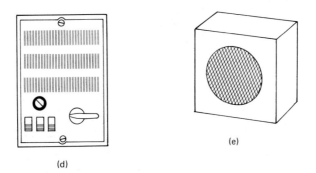

(d)

(e)

Fig. 2–7. Plans for interconnections of intercom wiring systems. (a) One master station and five remote stations. (b) Five master stations. (c) Four master stations and one remote station. (d) Four-station staff (semi-master) station. (e) High-power staff station for wall mounting.

utilized. The ground (*G*) wire is a common return circuit and the (*B*) wire serves to provide a circuit through the speaker when the function switch is in its *listen* position. In other words, this is a calling circuit. The station switch is arranged with six positions so that all master stations may participate in conference calls or any master station may select any one of the four other stations for private communication. If a receiver (earphone) is not used, the speaker serves both as a microphone and as a receiver.

A two-stage audio amplifier is employed in the arrangement of

Fig. 2–9, with input and output matching transformers. This is a utility amplifier and it has considerably less frequency response and more distortion than a high-fidelity amplifier. The amplifier operates at fixed gain, with a 25-ohm potentiometer in series with the ground-return wire for adjustment of the volume level. Note that the station selector switch provides for calling all of the five other stations simultaneously or for calling any one station only. To initiate a call, the receiver is lifted from its hook, and the *function* switch is thrown to its *talk* position. The *station* switch is then turned to the desired channel(s) and the caller speaks into the

(a)

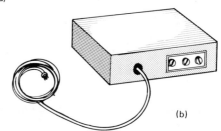

(b)

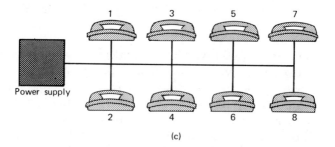

(c)

Fig. 2–8. Another type of intercom unit. (a) 16-station phone set. (b) Power supply. (c) Plan of an 8-station installation. (Courtesy of Bogen)

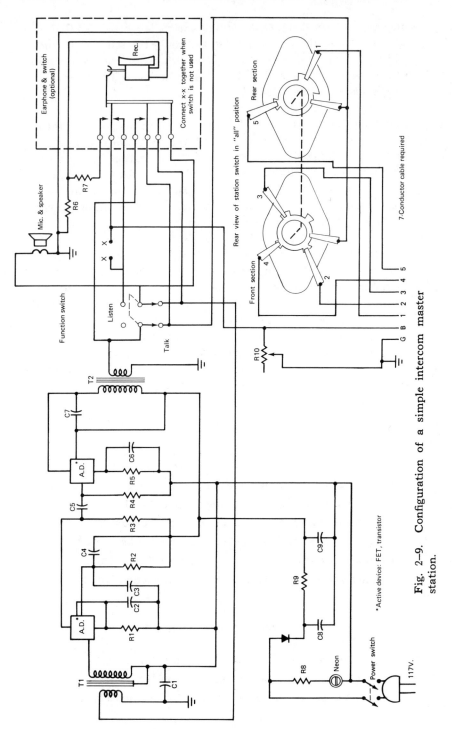

Fig. 2–9. Configuration of a simple intercom master station.

microphone (speaker). Observe that when the receiver is lifted from its hook, the B line is disabled, so that only the selected station(s) will hear the caller. A resistive pad is included in the receiver (earphone) circuit to provide a suitable audio level. The receiver can be eliminated, if desired.

Figure 2–10 shows a three-stage intercom configuration with switching facilities for an 18-unit system. This master unit may be utilized with other master units throughout, with a mixture of master and substation units, or with 17 substation units. It basically provides for individual calling only, although a conference call can be set up by calling in to the participants sequentially and instructing each to set his selector switch to the originating channel. Otherwise, this arrangement is essentially the same as explained for the configuration of Fig. 2–9. Instead of employing rotary switches, many intercoms are operated by push-button, as exemplified in Figure. 2–11. In this example, 13 stations may be included in the system. Three master stations, a selective substation, three substations with call in to the master stations, four confidential substations, and two substations without call in may be employed. The switches numbered from 1 to 6 serve to connect the master station to any of the other master stations or substations.

Figure 2–12 shows the cabling plan for the foregoing master stations and substations. The communication channels available are as follows:

No. 1 Master Station

Master switch No. 1 calls master station No. 2.

Master switch No. 2 calls master station No. 3.

Selecto-sub switch No. 3 calls selecto-sub station No. 1.

Confidential switch No. 4 calls confidential substation No. 1.

Regular substation switch No. 5 calls substation No. 2–1.

Regular substation switch No. 6 calls substation No. 3–1.

No. 2 Master Station

Master switch No. 1 calls master station No. 1.

Master switch No. 2 calls master station No. 3.

Selecto-sub switch No. 3 calls selecto-sub station No. 1.

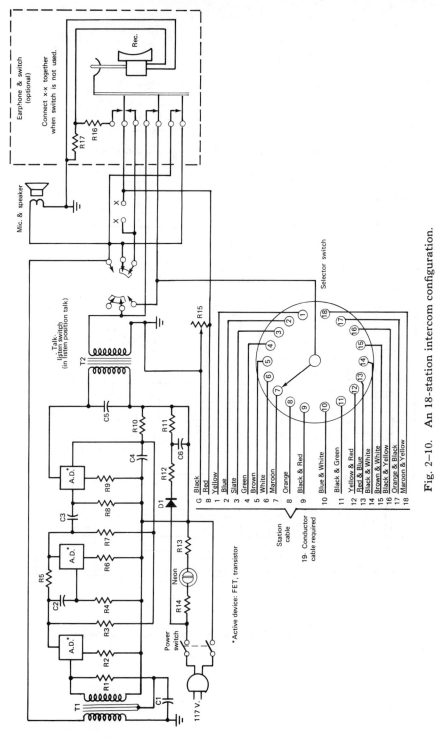

Fig. 2–10. An 18-station intercom configuration.

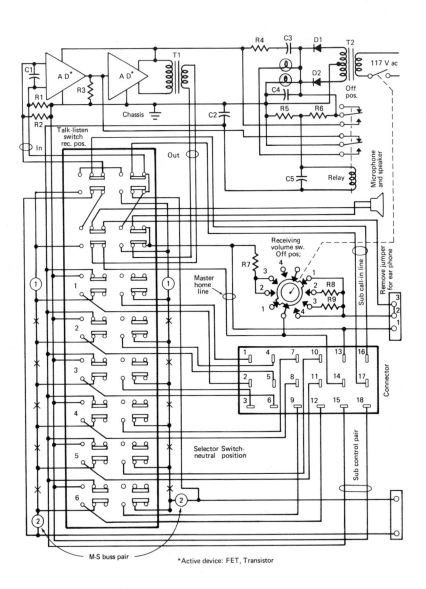

Fig. 2–11.   A push-button operated intercom configuration.

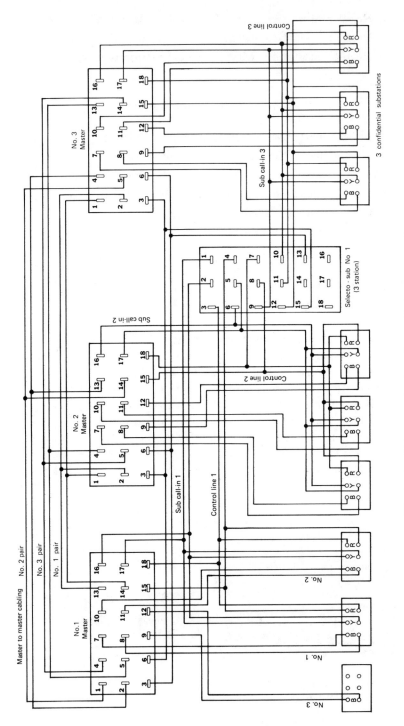

Fig. 2–12.   Master and substation cable interconnections.

Regular substation switch No. 4 calls substation No. 1–2.

Regular substation switch No. 5 calls substation No. 2–2.

Regular substation switch No. 6 calls substation No. 3–2.

No. 3 Master Station

Master switch No. 1 calls master station No. 1.

Master switch No. 2 calls master station No. 2.

Selecto-sub switch No. 3 calls selecto-sub station No. 1.

Confidential switch No. 4 calls confidential substation No. 1–3.

Confidential switch No. 5 calls confidential substation No. 2–3.

Confidential switch No. 6 calls confidential substation No. 3–3.

2.3   RADIO-INTERCOMMUNICATION SYSTEMS

Intercommunication systems may include a radio receiver for provision of radio programs to various rooms in a residence, shop, or factory. Figure 2–13 illustrates the units for a master station comprising three indoor remote stations and a door-entrance remote station. The indoor remote stations have latching-type talk/listen switches for continuous operation. An input jack for a record changer is provided for the master station. When a musical program is to be interrupted by an intercom call, an intercom override switch is thrown. Paging to all indoor remote stations is provided

Fig. 2–13.   An AM radio-intercom system. (Courtesy of Nutone)

by an all-call switch. Answers to paging calls can be made at remote stations without changing any control or switch settings. A door switch is provided at each station for speaking with a caller. This system provides AM radio reception only. More elaborate systems include FM radio reception and tape players.

FM-AM radio-intercom systems are generally designed for high-fidelity reproduction. Figure 2–14 shows a master station of this type with high-fidelity remote and utility remote units. Facilities are provided for room-to-room intercommunication and independent door-answering operation from any station. The switching arrangement permits turning off a radio program at one station for sending and receiving calls without interrupting the program at other stations. High sensitivity is provided so that remote stations can answer calls without walking up close to the unit. An eight-conductor cable is employed that permits installation of 16 remote stations. The hi-fi remote unit employs an 8-inch speaker, and the utility remote unit has a 5-inch speaker. These remote

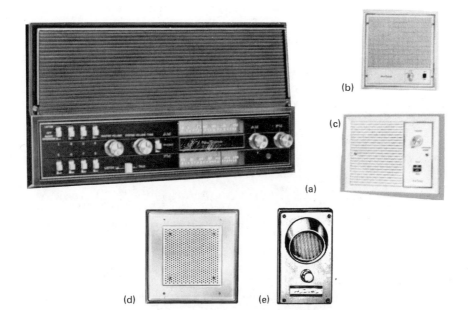

Fig. 2–14. An AM-FM radio-intercom system. (a) Master radio-intercom station. (b) High-fidelity remote station. (c) Utility remote station. (d) Door-answering remote station. (e) Outdoor substation with pushbutton for bell. (Courtesy of Nutone)

units are equipped with volume and function controls, and talk/ listen switches for indoor and door-answering stations. However, the door-answering unit has no switching facilities.

2.4 WIRING SYSTEM INSTALLATION

Installation instructions and color-coded wire are usually supplied by the manufacturer for intercom or intercom/hi-fi equipment. If comparatively long cable runs are required, additional cable must be procured. Figure 2–15 illustrates typical intercom cables of the unshielded and shielded types. Shielded cable is sometimes required, particularly in electrically *noisy* locations. The National Electrical Code stipulates that intercom wires or cables must not be run closer than two inches to any AC light or power conductors. This requirement applies to AC lines enclosed by conduit, in armored cable, in nonmetallic sheathed cable, and also to low-voltage AC lines. The code requirement ensures safety in case of electrical system malfunctions and also minimizes hum pickup by the intercom conductors. Similarly, intercom wires should not be run close to and parallel with telephone utility cables.

The National Electrical Code also requires that a lightning arrester be installed on each intercom line whenever there is any possibility of contact with a light or power line due to failure of a support or of insulation. (See Fig. 2–16.) Some intercom units are

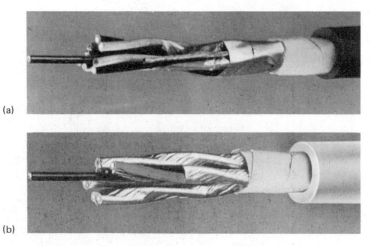

(a)

(b)

Fig. 2–15. Typical intercom cables. (a) Unshielded type. (b) Shielded type.

Fig. 2–16.   A lightning arrester. (Courtesy of G. C. Electronics)

in-wall types and others are on-wall types. A few intercom cases are designed for in-wall or on-wall mounting. A detachable frame is provided for this purpose. As noted previously, some intercom units are designed for placement on top of a desk. This type of installation is usually made with a connector and wall plate, such as illustrated in Fig. 2–17, both in the interest of appearance and so that the intercom unit can be unplugged from the wallplate, if desired. Elaborate master units are comparatively heavy and wall-mounted units must be securely attached to studs or to appropriate wood framing.

An intercom and music system is energized by a 120-volt line. Therefore, the installer must provide an outlet where the amplifier is located (usually at the master station). This is typically a per-

Fig. 2–17.   A wall outlet plate for intercom connection.

manent behind-the-wall outlet. However, on-wall types of intercoms may be powered from a convenience outlet, particularly in
old work (installation in a finished building). Concealed wiring is
utilized when it is practical. However, brick or equivalent construction in old work necessitates exposed wiring which may be run in
surface metal raceways, both for mechanical protection and for
shielding against possible noise and/or hum pickup. Figure 2–18
shows typical metal raceways. In new work (installation while a
building is under construction), junction boxes are usually installed
in the same manner as for light or power wiring. Note that approximately three inches of intercom cable should be left hanging out of
the box for final connections, pending building completion.

Figure 2–19 illustrates how a junction box is nailed to a stud.
The edge of the box must project forward sufficiently to be even
with the panelboard, sheetrock, building board, or plaster surface
of the finished wall. Cables are run through holes in studs and
plates, as exemplified in Fig. 2–20. However, in old work, studs cannot be drilled, and concealed wiring must be *fished* behind walls as
dictated by the building construction. Figure 2–21 shows the steps
that are generally involved in *fishing* a wire or cable. In this example, the cable is run from one location to another through an attic
or upper room. In much the same manner, a cable may be run
under the floor across a basement, as exemplified in Fig. 2–22.

In new work where extensive runs must be made in electrically
noisy locations such as factories or high-rise buildings in metropolitan areas, it is good practice to install an intercom *pipe job*.
This denotes the use of electrical conduit or EMT (thin-wall conduit) to enclose the intercom cable(s) over their entire route. Conduit is always grounded to a cold-water pipe and it provides highly

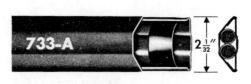

Fig. 2–18.   Typical surface metal raceway.

Fig. 2–19. How a junction box is nailed to a stud.

Fig. 2–20. Cables are run through holes in studs and plates.

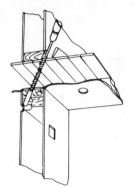

If there is access into attic or upper room, remove the upstairs baseboard. Then drill a diagonal hole downward as shown.

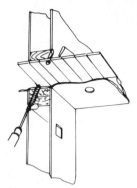

Drill a diagonal hole upward from opposite room. Then drill horizontally until holes meet. This method requires patching of the plaster.

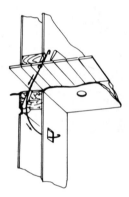

Push a 12-foot fish wire, hooked at each end, through hole on second floor. Pull one end of fish wire through swith outlet on first floor.

Next, push a 25-foot fish wire, hooked at each end, through ceiling outlet. Then fish until the first wire is caught.

Next, withdraw either wire until it hooks the other wire; then withdraw the second wire until both are hooked together.

Finally, pull the shorter wire thru the switch outlet. When long wire hook appears, attach the cable and pull it through the wall and ceiling.

Fig. 2—21.   How to "fish" a wire or cable.

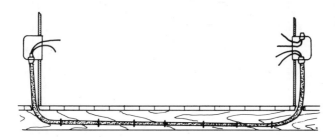

Fig. 2–22.   Cable routed through floor and across basement.

efficient shielding. No other conductors should be installed in the same conduit with intercom wires or cable. In addition to shielding, conduit also provides good mechanical protection for cables. The chief disadvantage of a conduit installation is its comparatively high cost.

2.5   TROUBLESHOOTING INTERCOM SYSTEMS

Trouble symptoms in intercommunication systems include completely dead system, partially dead system, weak signal(s), distorted output, and intermittent operation. When the system is completely dead, and only one master station is present, the power supply (Fig. 2–23) falls under suspicion first. On the other hand, if the system includes two or more master stations, a cable break is often responsible. However, when the system is partially dead and two or more master stations are employed, a defective power supply is a logical suspect. Weak signal strength can be caused by more than one type of fault. For example, if all the outputs from a master station are weak, the amplifier is most likely to be defective. However, if one or two outputs only are weak, the trouble is likely to be found in the switching system. Similarly, if incoming calls are weak, but outgoing calls are normal, a preliminary localization of the trouble area becomes apparent.

In practically all trouble situations, a schematic diagram of the intercom system is indispensable. Most schematics indicate component values, and many specify normal operating voltages. Amplifier troubleshooting involves the same basic approaches and

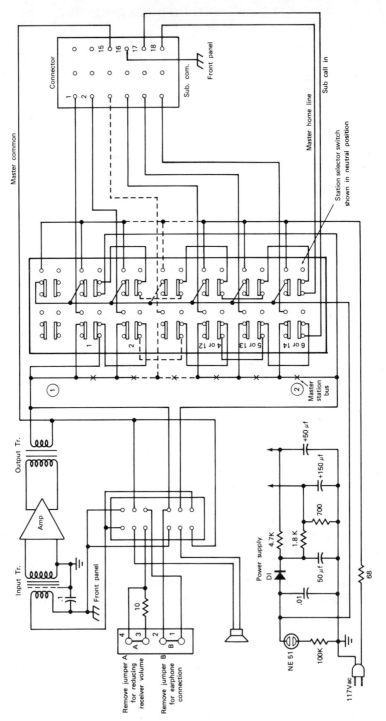

Fig. 2-23. Typical intercom switching system.

procedures that were detailed in the previous chapter. Note that even when all operating voltages are not specified, useful clues can be obtained from one or two specified voltages. For example, refer to Fig. 2–23. The power supply has three output branches. Only the voltage across $C2$ ($+7.1$ volt) is specified. Nevertheless, if we measure this voltage value across $C2$, we may conclude that the power supply is basically normal, and that we need only check $C3$ for possible leakage or a short-circuit. Note that if $C3$ is leaking badly, $R1$ will be very hot. In case $C3$ becomes short-circuited, $R1$ is quite likely to burn out.

Defects in the switching system are usually the result of poor contacts. A poor contact can develop as the result of wear, corrosion, excessive current flow due to defects elsewhere in the system, or mechanical damage. Switch trouble is comparatively common in humid areas or in areas subjected to fumes or vapors at times. Contact resistance can be checked by closing the switch contacts and measuring the contact resistance with an ohmmeter. A milliohmmeter such as illustrated in Fig. 2–24 is more informative than a

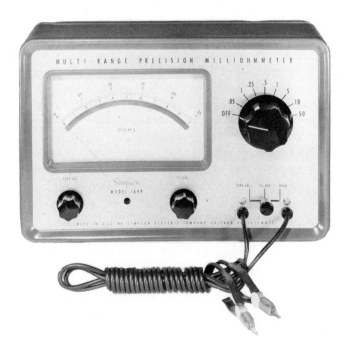

Fig. 2–24.   A milliohmmeter used to check switch contact resistance. (Courtesy of Simpson Electric Co.)

conventional ohmmeter in pinpointing defective switch contacts. It is advisable to tap the switch knob or lever while measuring contact resistance. Occasionally, this check will disclose an intermittent condition. Contacts can sometimes be restored to normal operation by cleaning with a spray such as used in TV tuner service. Otherwise, the defective switch must be replaced.

Cable conductors can be checked in pairs for continuity with an ohmmeter. In other words, a pair of conductors are short-circuited together at one end with clip leads and tested at the other end for continuity with an ohmmeter. Although most cable faults involve open circuits, trouble is occasionally caused by short-circuited conductors. This possibility can be checked by throwing all the associated switches to their open positions and checking the suspected pair of conductors with an ohmmeter. If there is a short-circuit between the conductors, the ohmmeter will read the loop resistance of the wires (normally a comparatively low value). Or, if there is substantial leakage between the conductors, the ohmmeter will read higher than the resistance value of the wires, but considerably less than infinity. On the other hand, if the ohmmeter reads infinity, the wires are in normal condition (assuming that they have previously passed the short-circuit test).

When intercom equipment appears to be in normal operating condition and cable defects are not found, it is advisable to check the associated connectors. Sometimes a multicontact plug develops a corroded or otherwise defective pin and fails to make proper contact in the receptacle. In equipment that employs an external microphone with a cord, or an earphone with a cord, the cords should be checked for continuity. The microphone output can be checked directly with an audio voltmeter, if this instrument is available. (An audio voltmeter provides a very sensitive indication on its first range.) If intermittent operation is encountered, troubleshooting is practical only during the inoperative interval. Some times an intermittent can be speeded up by tapping the components, by switching the equipment off and on, or by heating the pigtails of suspected components. (Semiconductor components in general should not be heated, as they are susceptible to thermal damage.)

Keep in mind that comparison tests can often be made in an intercom system, in case complete technical data are unavailable. In other words, suppose that there are two or three master units in the system, one of which has been identified as defective in operating tests. The circuits and component values in the defective unit can be checked against voltage and resistance values measured in one of the normal units. In difficult circuit problems, comparison tests can be helpful to supplement complete technical data. For instance, technical data never provides stage-gain values and audio levels through the system. When this information is needed, it can be obtained directly from a normally operating unit with an audio voltmeter. This type of test work requires a small audio oscillator to provide a steady tone at the input end of the system.

3

PUBLIC-ADDRESS SYSTEMS

3.1   GENERAL CONSIDERATIONS

A public-address system consists basically of a microphone, amplifier, and speaker(s) to facilitate the communication of intelligible speech to groups of people. Coverage of large groups may be the chief purpose of providing communication over outdoor areas or of providing sound reinforcement indoors with sufficient fidelity that the sound system is unobtrusive. Acoustic problems differ considerably for outdoor and for indoor installations. Directional horns are often employed in outdoor systems to concentrate sound energy efficiently over the areas served. (See Fig. 3–1.) Intelligibility is the usual performance criterion and the system frequency response is typically limited to a range of 300 Hz to 5,000 Hz. Intelligibility may also be the performance criterion for an indoor system such as an airport waiting room. On the other hand, high-fidelity reproduction is required for a sound-reinforcement system, for instance, a dance or concert hall.

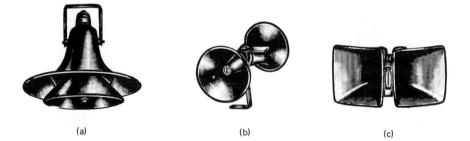

(a)   (b)   (c)

Fig. 3–1. Directional horns used in outdoor PA systems.
(a) Radial reflex type, 360° horizontal dispersion. (b)
Bidirectional projector. (c) Cobreflex type, with wide dispersion.

Outdoor installations utilizing directional horns require approximately 2 watts of audio power per 1,000 square feet and operate at efficiencies up to 50 percent. Sound reflection from stadium walls, grandstands, fences, or buildings may increase the sound level, but often at the expense of intelligibility. Therefore, the acoustics of the served area are often a dominant consideration in system design. Reliability is often a basic requirement, necessitating more or less duplication of facilities, arranged so that service will not be interrupted in the event of some equipment failure. In other words, if one of the high-power amplifiers fails, it is desirable that service continue, even at a reduced sound level. Similarly, if a speaker or line should fail, it is desirable that the remaining speakers and lines continue to operate normally.

Indoor sound-reinforcement systems generally employ wide-range speakers. A minimum audio power of 50 watts is required for seating capacities up to 500, and an additional 50 watts for capacities up to 1,000, and so on. It is good practice for the control operator to be stationed in a room communicating with the served area, so that audio levels can be controlled in accordance with realistic requirements, and maintained at a level sufficient to override prevailing audience noise. Installations operating at comparatively high audio power levels, either outdoors or indoors, have been standardized in accordance with the 70.7-volt speaker-matching system established by the Electronic Industries Association (EIA). This system reduces losses in the wiring, and increases operating efficiency, as explained in greater detail subsequently.

3.2 PLANNING A PUBLIC-ADDRESS SYSTEM

As a general rule, public-address system planning starts with calculation of the area to be served, of required audio power, and of type(s) of speakers to be utilized. The approximate number of speakers is estimated next. As an illustration, most office areas can be served satisfactorily with 8-inch cone speakers rated at $\frac{1}{2}$ watt each. Ceiling speakers are often installed about 25 feet apart. Warehouses or meeting halls often require 5-watt speakers mounted from 25 to 50 feet apart. When noise is troublesome, 5-watt indoor horn speakers may be installed at necessary locations to maintain a satisfactory audio-to-noise ratio. Speakers serving spectator groups in outdoor installations should be placed near the audience. For distant coverage up to $\frac{1}{2}$ mile, clustered projectors are employed, as detailed later. Roller rinks and school auditoriums are often best served by sound columns, comprised of several speakers mounted in the same vertical or horizontal housing. (See Fig. 3–2.)

After the required audio power has been determined by adding the power ratings of the speakers in the system, a suitable am-

Fig. 3–2. A sound column, comprising several woofers and tweeters. (Courtesy of Heath Co.)

plifier is chosen. It is good practice to operate an amplifier below its maximum-rated output level. For example, if 6 horn-type speakers rated at 6 watts each and 5 cone-type speakers rated at 3 watts each are used, the total audio power requirement is 51 watts. In turn, an 80-watt amplifier would be a conservative choice, permitting the addition of a few special speakers in the event that they may be needed. Most **PA** amplifiers have provisions for one or more microphones and one or more other sound sources such as **FM-AM** radio tuner outputs, **TV** sound output, tape-recorder, and record-player outputs. Level (fader) controls may be provided to adjust the volume from each of several sources simultaneously. High-impedance inputs are generally provided and matching transformers can be employed to accommodate low-impedance sound sources. Figure 3–3 shows the appearance of typical amplifiers.

Most **PA** amplifiers have speaker outputs for a constant-voltage system (either 25 or 70.7 volts) plus 4, 8, and 16 ohm out-

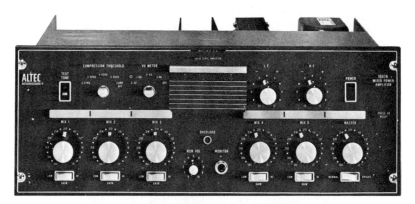

(a)

(b)

Fig. 3–3.   Typical public-address amplifiers. (a) 80-watt silicon transistor amplifier (Courtesy of Altec Corp.). (b) 200-watt amplifier (Courtesy of Bogen, Inc.).

puts to match voice coils directly. A 25-volt system is easier to install because conduit is not required. On the other hand, a 70.7-volt system provides comparatively high operating efficiency. After the amplifier has been selected, suitable microphones should be considered. The choice of microphone is chiefly determined by intended use. In other words, no one microphone will perform equally well under all conditions. The basic considerations are the pickup element that is utilized, the pickup pattern that is optimum for the location, and the conditions of microphone use. Ceramic microphones are comparatively inexpensive, have relatively high output and wide frequency response, and resist climatic changes. Dynamic microphones resist shock and bad weather conditions, have superior frequency response and fidelity, and are well suited for orchestral or vocal reproduction. Carbon microphones are inexpensive and rugged, have high sensitivity, but also have the disadvantages of inferior frequency response and comparatively high distortion. Figure 3–4 illustrates typical public-address types of microphones.

The pickup pattern of a microphone is determined by the design of its housing. Various patterns are omnidirectional, bidirectional, cardioid, and differential. An omnidirectional microphone picks up sound energy from all directions. A bidirectional microphone responds to sound from front and back directions. A cardioid microphone picks up sound from the front only, while differential

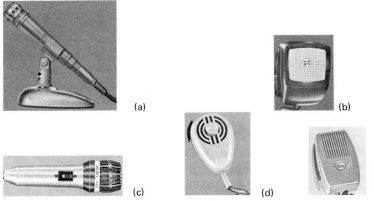

(a)   (b)   (c)   (d)   (e)

Fig. 3–4. Typical public-address microphones. (a) Omnidirectional ceramic microphone. (b) Cardioid ceramic microphone. (c) Unidirectional dynamic microphone. (d) Noise-canceling dynamic microphone. (e) Noise-canceling carbon microphone. (Courtesy of Turner Co., Inc.)

(noise-canceling) microphones do not respond to sound waves unless they originate a few inches from the diaphragm. The majority of microphones have high output impedance. If they are to be used with cables over 40 feet long, a matching transformer is required to maintain good frequency response. Microphones with low output impedance can be used directly with long cables.

Ceramic and dynamic microphones are generally recommended for voice and music reproduction. High-quality dynamic microphones, sometimes called broadcast-quality microphones, have high-fidelity response and are commonly used with PA systems for auditoriums, stage groups, theaters, and concert halls. Carbon microphones are utility-type transducers and are used only in systems where economy is desired and high-quality reproduction is not necessary. Comparatively, microphones generally have substantially better characteristics than speakers. Microphones are also affected less by poor acoustics than are speakers. In many PA systems, a wide choice of microphones is permissible, whereas speaker type, placement, and energization may be critical and demanding.

3.3   THE 70.7 AND 25 VOLT SYSTEMS

As noted previously, the 70.7 and 25 volt speaker-matching systems permit higher operating efficiency in audio power systems. The 70.7 volt system is preferred for high-power PA networks. Figure 3–5 depicts a 70.7 volt matching network for three speakers. A matching transformer is installed to match a speaker, or a group of speakers, to the 70.7 volt line. This is called a *constant-voltage system*, because the line voltage is comparatively unaffected by switching speakers on or off. Network calculations are as follows:

1. Determine the power rating of each speaker.
2. Add the power values to find the total power demand and use a 70.7-volt amplifier with a rated power output at least equal to this demand.
3. Select a 70.7-volt matching transformer for each speaker (or for each group of speakers) with appropriate primary wattage ratings.
4. Connect the primary terminals of each transformer across the 70.7-volt line from the amplifier output. Note that a primary mismatch up to 25 percent is tolerable.

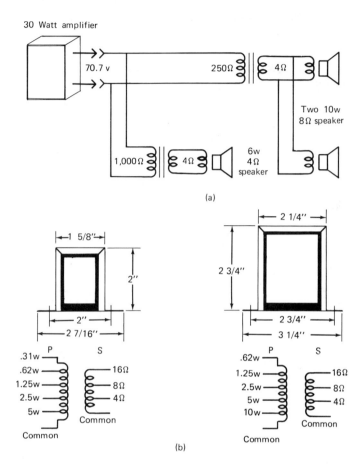

Fig. 3–5.   Example of a 70.7-volt speaker-matching network. (a) Configuration. (b) Typical 70.7-volt transformer taps and dimensions.

5. Connect the secondary terminals of each transformer to its speaker (or group of speakers), observing the matching ohms tap.

6. In case the matching transformers might be rated in impedance values, the primary wattage of a transformer may be calculated as follows:

$$z_p = \frac{70.7^2}{p}$$

where z_p is the rated primary impedance and P is the wattage rating of the speaker.

It follows that the following power and impedance relations result:

1 watt corresponds to 5,000 ohms z_p

2 watts correspond to 2,500 ohms z_p

5 watts correspond to 1,000 ohms z_p

10 watts correspond to 500 ohms z_p

Refer to Fig. 3–5. The 6-watt speaker has a 4-ohm voice coil, and the paralleled 10-watt speakers have 8-ohm voice coils. In turn, the total power demand is 26 watts and the amplifier would be rated for approximately 30 watts output. Using the above equation, the primary impedance for the 6-watt transformer would be 833 ohms; a 1,000-ohm impedance could be utilized. For the 20-watt speaker combination, the primary impedance would be 250 ohms.

3.4   MICROPHONE CONSTRUCTION

The single-button type of carbon microphone was described previously. In addition to this basic type, the double-button type of carbon microphone depicted in **Fig. 3–6** is also used in PA systems. The double-button construction has an advantage of lower distortion, due to its push-pull action which cancels even harmonics. As shown in the figure, a double-button carbon microphone has carbon-granule cups on both sides of the diaphragm. The matching transformer has a center-tapped primary winding. Jacks are provided for plugging a milliammeter into the circuit. The meter indicates the average DC current flow, which is adjusted by means of the rheostat. Double-button microphones generally have a total resistance of several hundred ohms. On the other hand, the gate-input impedance of a field-effect transistor, for example, is extremely high. Accordingly, exact impedance matching is not practical; the matching transformer is often designed to match several hundred ohms to 500,000 ohms.

Operation of a double-button carbon microphone is shown in **Fig. 3–7**. The output waveform from each button is distorted, principally by generation of a second harmonic. In turn, the push-pull

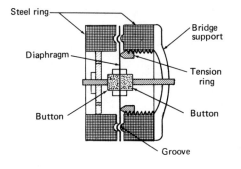

(a)

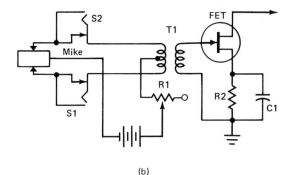

(b)

Fig. 3–6. (a) Double-button carbon microphone. (b) Input circuit.

matching transformer combines the pair of waveforms as shown, thereby providing cancellation of the second harmonic. Note that push-pull action cancels even harmonics only; it enhances odd harmonics. However, odd-harmonic distortion from a double-button microphone is very small compared to even-harmonic distortion. The maximum rated DC current for a carbon microphone should not be exceeded because the tiny arcs between carbon granules develop heat and some of the granules may be fused. In turn, the sensitivity of the microphone is reduced and the granules are said to be *packed*. This difficulty can be overcome by reducing the DC current flow and tapping the diaphragm lightly to unpack the granules. Note that a high-quality single-button carbon microphone will provide about 0.25 AC volt across 75 ohms, whereas a high-quality double-button carbon microphone will provide about

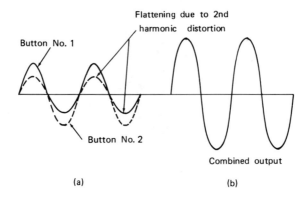

Fig. 3–7. Operation of a double-button carbon microphone. (a) Outputs from individual buttons. (b) Push-
pull output.

0.05 AC volt across 200 ohms. The rated DC current flow is typically 25 mA for each button.

Ceramic and crystal microphones are also used in PA systems.
Figure 3–8 shows the input circuit for a piezoelectric or crystal type
of microphone coupled with an amplifier employing a field-effect
transistor. In this example, a pair of Rochelle-salt crystals (potas-
sium-sodium tartrate) are cemented together; electroplated metal
films provide contact to opposite sides of the unit. In one type of
construction, sound waves strike one of these metal films, causing
the crystal unit to vibrate. In turn, the piezoelectric effect of the
crystal lattice generates an audio output voltage. Another type of
construction employs a diaphragm that is mechanically coupled to
the crystal unit. This arrangement provides a greater output voltage, although it has poorer frequency response than the crystal
unit alone.

In some microphones, several crystal units are connected in
series, parallel, or series-parallel to obtain higher output voltage,
lower internal impedance, or both. A crystal microphone is basically
a generator and does not use a battery. Note that if substantial DC
voltage were accidentally applied to a crystal microphone, the unit
would be damaged. Although the output level of various crystal
microphones differs appreciably, depending on the design, an output of 0.02 volt AC is typical. Since a crystal microphone has a
very high internal impedance (typically 250,000 ohms), shielded
cable must be used. If the cable is more than a few feet in length, a

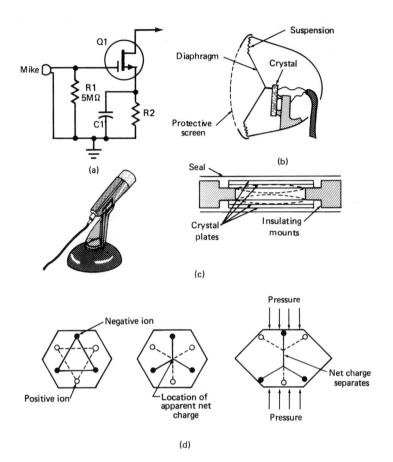

Fig. 3–8. (a) Crystal microphone circuit. (b) Plan of crystal microphone. (c) Appearance of a crystal microphone. (d) Generation of voltage by a Rochelle-salt crystal. (Courtesy of Shure Brothers, Inc.)

matching transformer or device must be inserted between the microphone and the cable to maintain adequate frequency response.

Figure 3–8(d) provides a visualization of voltage generation by a Rochelle-salt crystal when subjected to mechanical pressure. The crystal lattice is an ionic structure that is the basis of the piezoelectric effect. When pressure is applied to the crystal, a displacement of charge centers occurs in the lattice. The crystal is made up of equal numbers of positive and negative ions arranged in equilateral triangles. In turn, the average or net charge in group of three like charges is in the geometric center of this triangle. We observe that the centers of net negative and positive charges coincide when no pressure is applied to the crystal.

When pressure is applied, the ions in the lattice are forced to move. This movement results when each triangle of charges is compressed in a vertical direction and each triangle is spread out in a lateral direction. As the shape of each triangle changes, the location of the apparent net charge also changes. The net positive charge moves upward and the net negative charge moves downward. Because of this charge separation, a voltage is developed within the crystal. The individual charge separations produce a combined voltage difference that appears on the surfaces of the crystal. In this illustration, the top surface of the crystal exhibits a positive potential and the bottom surface exhibits a negative potential.

Velocity microphones are also utilized in PA systems. Figure 3–9 depicts the plan of a velocity or ribbon-type microphone. A thin corrugated metal ribbon is suspended in a strong magnetic field. Thus, the microphone utilizes generator action; a voltage is induced in the ribbon when it vibrates in response to incident sound waves. We will recognize that the internal impedance of a velocity microphone is very low; therefore, a microphone transformer is generally used to step up the microphone impedance and match it to the characteristic impedance of the cable. In turn, a line transformer is used to match the cable impedance to the high impedance of the amplifier input circuit. In Fig. 3–9(b), an audio signal amplitude of approximately 0.04 volt AC is applied to the gate. To obtain good frequency response over the audio range, the mass of the ribbon in the microphone is chosen to provide a mechanical resonant frequency below the range of ordinary speech and music waveforms.

3.5   PA AMPLIFIERS

A PA amplifier may be designed as a unit, or it may consist of separate amplifiers, as exemplified in Fig. 3–10. When more than one audio input source is utilized (such as two microphones), operating controls are provided to change from one source to the other or to mix the outputs of the sources. The mixer and fader section is designed to provide this flexibility of operation without sudden increases or decreases in volume. A fader serves to gradually

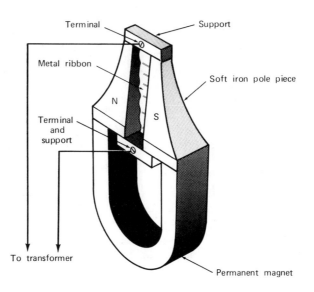

Terminal
Support
Metal ribbon
Soft iron pole piece
N
S
Terminal
and
support
To transformer
Permanent magnet

(a)

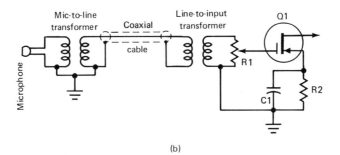

Mic-to-line
transformer
Coaxial
Line-to-input
transformer
Q1
cable
Microphone
R1
R2
C1

(b)

Fig. 3–9.   (a) Plan of a velocity (ribbon) microphone.
(b) Input circuit.

reduce the volume of one source while gradually increasing the volume of another source. A mixer may or may not provide fader action. As an illustration, Fig. 3–11 shows a fader network that does not provide mixer action and a mixer device with controls for fading the sources in or out.

A microphone amplifier provides preliminary step-up of the microphone output signal. Figure 3–12 shows a configuration for a three-stage amplifier designed for use with high-level microphones.

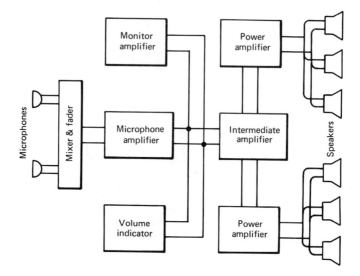

Fig. 3–10.   Block diagram of a public-address system.

It provides a total voltage gain of 1500 to 2000 times with a dynamic range of 5 volts rms output across a load of 500 ohms from an input of 0.4 volt rms. The frequency response is essentially flat from 20 Hz to 30 kHz. DC power requirements of the amplifier are 20 volts at 30 milliamperes. A low-noise transistor is used in the first stage which operates in class *A*. The second and third stages

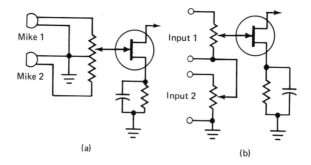

Fig. 3–11.   Amplifier input arrangements for two audio sources. (a) A fader circuit that does not provide mixer action. (b) A mixer device with fader controls.

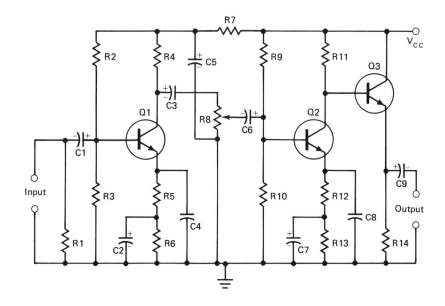

Fig. 3–12. A microphone preamplifier configuration. (Courtesy of RCA)

comprise a direct-coupled class *A* driver stage and an emitter-follower output stage. Either low-impedance or high-impedance microphones can be employed, provided that the value of the input resistor $R1$ is selected to match the microphone line impedance up to a maximum of 10,000 ohms.

Monitor amplifiers are typically used as line amplifiers in which the monitor position (usually with headphones) is located at an appreciable distance from the signal source. Figure 3–13 shows a circuit for a typical monitor amplifier. It provides a flat frequency response from 20 to 20,000 Hz and is suitable for driving any line that has an impedance of 250 ohms or more. The amplifier operates from a DC supply of 12 volts and can provide a maximum undistorted output of 3 volts rms into a 250-ohm line. The voltage gain and input impedance of the amplifier are determined by the values of the emitter resistor $R3$ and the feedback resistor $R4$. A tabulation in Fig. 3–13 gives values of these resistors for various voltage gains from unity to 166 and for input impedances from 2700 ohms to 55,000 ohms. Note that a volume indicator is a special type of dB meter that indicates voice or music levels in volume

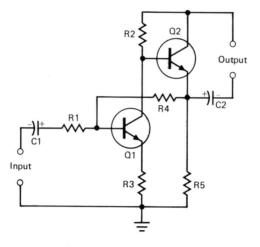

Resistance data for different voltage gains and input impedances, for an output of 1 volt rms into a 250-ohm line.

Voltage gain	Input impedance (ohms)	R3 (ohms)	R4 (kilohms)
166	2700	0	680
22	7300	39	470
17	9000	68	430
10	15000	100	390
3	55000	390	360
1	100000	1200	330

Fig. 3–13. Typical monitor amplifier circuit. (Courtesy of RCA)

units (VU). It is designed to respond in an optimum manner to complex waveforms, instead of sine waveforms, as do dB meters.

The power amplifiers in Fig. 3–10 are driven by an intermediate amplifier, which is basically similar to a microphone amplifier, except that it may employ higher power transistors. If a power amplifier has sufficient gain, it may be driven directly by the microphone amplifier. Figure 3–14 shows a configuration for a high-quality 70-watt audio power amplifier. It has a frequency response from 5 to 25,000 Hz, with less than 1 percent harmonic distortion. No driver or output transformer is utilized. The driver stage employs a pnp and an npn transistor connected in complementary symmetry to develop push-pull drive for the output stage. Silicon transistors are used in the output stage, connected in series

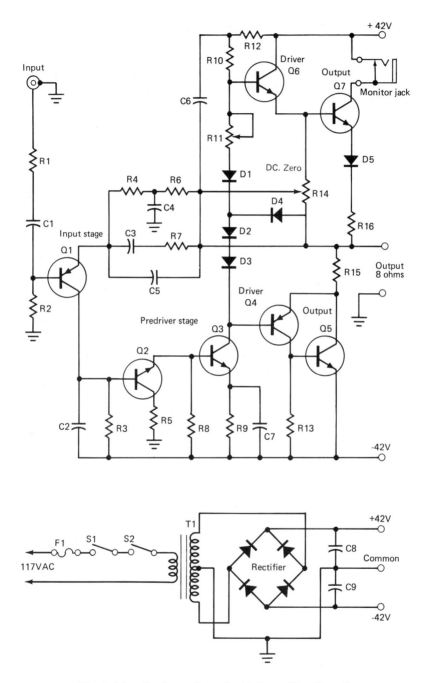

Fig. 3–14.   Configuration of a high-quality 70-watt audio power amplifier. (Courtesy of RCA)

with the positive and negative supply voltages. Negative feedback of 35 dB is provided by $R5$ and $C3$. The output is directly coupled with an 8-ohm speaker.

The input stage of Fig. 3–14 operates in the common-emitter mode and provides DC feedback through $C4$, $R4$, $R6$, and $R14$ (the DC zero adjustment) to maintain the quiescent voltage of the output stage at zero volts. The predriver stage operates as a Darlington pair. It has a minimum loading effect on the input stage and provides substantial voltage amplification. Note that the driver and output stages provide power gain, but do not develop voltage gain. Bias voltage is adjusted by $R11$ for the complementary driver stages and is regulated by three $1N3754$ diodes. $R11$ is adjusted for 20 milliamperes at monitor jack J. The output stages are operated in class AB to avoid crossover distortion.

Short-circuit protection for the amplifier in Fig. 3–14 is provided by a current-limiting circuit consisting of a zener diode and emitter resistors $R15$ and $R16$. If a current of more than five amperes flows through either resistor for any reason, the voltage across the zener diode will cause forward conduction during the negative-going output half-cycle and cause it to break down at the diode reference voltage during the positive-going output half-cycle. The driving voltage, therefore, is clamped at that level and any further increase in output current is prevented. Thereby, both the output transistors and the driver transistors are protected against accidental overloads. In case of a sustained overload, the thermal cutout switch $S2$ will turn off the power to the amplifier.

3.6   PA SYSTEM CHARACTERISTICS

In speech reinforcement systems, the listener hears sound from two sources. In other words, part of the sound arrives from the person making the speech and another part of the sound arrives from the PA system. The total sound that arrives is greater than that from either source alone, but the two portions do not arrive simultaneously. From a practical viewpoint, the maximum tolerable time delay between the two arriving wavefronts is approximately 25 milliseconds. A longer time delay is likely to result in a confusing *separation* of the two sound sources; the listener's attention jumps from one source to the other and becomes distracted. This time-delay effect is closely related to the echo effect. Since the human ear has a natural tolerance for echo effects, an echo tends to be disregarded under ordinary conditions. However, if the echo arrives too long after the original wavefront, it is no longer disregarded and the listener becomes uncomfortably aware of the *split* sound sources.

Under ordinary conditions, an echo is weaker than the true sound source. In turn, the human ear has less tolerance for echoes that are stronger than the true sound source. In a speech-reinforcement system, 10 percent of the listeners will be disturbed if the PA system level is 3 dB below the true source and is delayed 60 milliseconds. Again, 10 percent of the listeners will be disturbed if the PA system is 6 dB below the true source and is delayed 80 milliseconds. On the other hand, 10 percent of the listeners will be disturbed if the PA system is 10 dB above the true source and is delayed 30 milliseconds. Therefore, it is good practice to plan a PA system so that no listener will be exposed to two arriving wavefronts that have a time delay greater than approximately 25 milliseconds. This is called the Haas effect, summarized in Fig. 3–15.

A cluster of speakers or a speaker column will provide considerable sound coverage. However, in large buildings that have appreciable reverberation, a large speaker column may need to be supplemented by smaller columns placed farther back from the sound source. For example, a large speaker column may be installed near the rostrum which might cover a distance as great as 100 feet. Then, a supplementary speaker column can be installed with a time-delay circuit so that the sound from the first column arrives

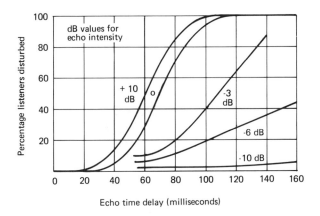

Fig. 3–15.   Time-delay disturbance versus intensity of reflected sound.

5 milliseconds before the sound from the second column is radiated. After another 50 feet, for instance, a second supplementary column may be installed with a time delay so that to a nearby listener the sounds arrive first from the main column, 5 milliseconds later from the first supplementary column, and still another 5 milliseconds later from the second supplementary column. However, the intensity of the delayed sound should not be more than 10 dB greater than that of the first-arriving sound. Otherwise, separation is likely to become evident.

Time-delay units for **PA** systems generally employ a tape deck with spaced playback recording and playback heads, or a magnetic disk. (See **Fig. 3–16.**) When delay systems are installed, it is im-

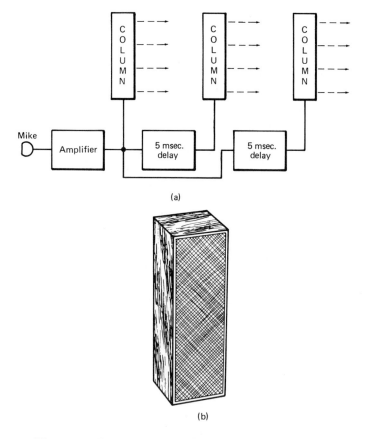

Fig. 3–16.   PA system with time-delay units. (a) System arrangement. (b) A sound column. (Courtesy of Allied Radio Co.)

portant to limit the amount of sound energy that travels back toward the front of the building or area. It is necessary that the rearward radiation from a delayed source be at least 6 dB less than the forward radiation. One way of reducing the rearward radiation is to put a layer of cotton 2 inches thick at the back of the delay column. The cotton layer changes the phase of the sound radiated from the back of the speaker by an amount equal to the phase change of the sound in traveling from the front around the column to the back. The effect is to reduce the radiation of rearward sound appreciably. In difficult locations, it is helpful to reduce the frequency response of the PA system to a range of 250 to 4,000 Hz.

3.7   PORTABLE MEGAPHONES

Portable electronic megaphones such as illustrated in **Fig. 3–17** are used by construction workers, law-enforcement officers, boat navigators, and so on. The unit comprises a reentrant horn speaker with a microphone mounted on the back. A 5-transistor class *B* amplifier is typically employed to step up the microphone signal. The unit is powered with a 9-volt battery and has a maximum output of 8 watts with a sound range of 600 yards. A volume control is provided and a built-in siren signal source is available in the illustrated unit. Another design is shown in Fig. 3–18. This is a 25-watt unit that weighs 13 pounds (more than twice as much as the portable megaphone). A switch is provided to change the circuit configuration into a sensitive directional listening arrangement. An input jack is available for feeding in a signal from a record player, tape player, or radio tuner.

Fig. 3–17.   A portable electronic megaphone. (Courtesy of Fanon/Courier Corp.)

Fig. 3–18. A 20-watt PA talk-listen portable unit. (Courtesy of Fanon/Courier Corp.)

3.8 PAGING ARRANGEMENTS

Relays such as shown in Fig. 3–19 are used to tie-in plant PA systems to paging microphones with provision for talkback from a horn or speaker. In other words, the horn operates as a microphone during the talkback period in the same basic manner as the horn unit in Fig. 3–18. A relay also makes it possible to tie-in an intercom system with a PA system for paging through a plant. Switching is automatic at the output end, and the person paged merely replies in the direction of the horn.

Fig. 3–19. Relay employed with PA system for paging/ talkback operation. (Courtesy of Potter & Brumfield, Div. of AMF, Inc.)

4

BROADCASTING STUDIO AND FIELD AUDIO SYSTEMS

4.1   GENERAL CONSIDERATIONS

Broadcasting audio systems comprise transducers, control devices and units, transmission means, amplifiers, monitoring equipment, and specialized signal-processing equipment. Acoustics are also a part of the system. In turn, audio engineers and technicians are concerned with various types of microphones, preamplifiers, microphone mixers, studio amplifiers, limiting (compression) amplifiers, volume controls and faders, volume indicators, monitoring facilities, intercommunication line facilities, relays and switching apparatus, telephone line facilities, equalizers, and studio acoustics. Several studios may be utilized, and remote pickups (*nemo* programs) contribute to the comparative complexity of broadcast audio systems. Figure 4–1 shows an example of broadcast studio and field equipment. It is evident that a large part of the system is concerned with audio signal switching and control functions.

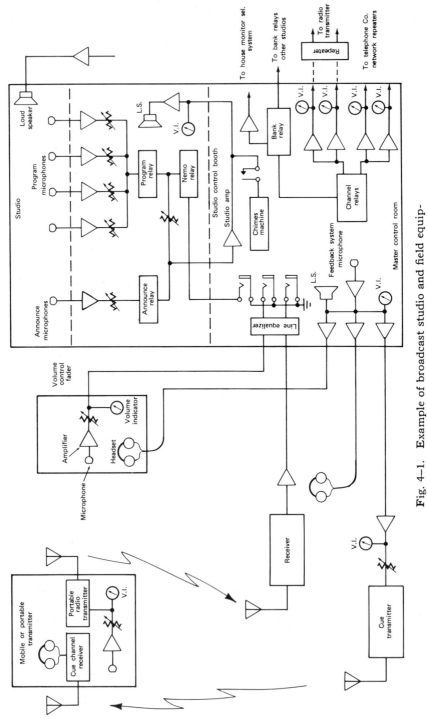

Fig. 4-1. Example of broadcast studio and field equipment.

Microphones are of the velocity, dynamic (moving-coil), crystal, and capacitor types. Unidirectional, bidirectional, and omnidirectional (nondirectional) types are used. Polydirectional microphones provide a choice of directional response. They typically have ribbon (velocity) elements in elaborated housings that provide desired responses to incident sound waves. The desired response is controlled by means of a movable shutter. In addition, the low-frequency response of the polydirectional microphone is adjustable. A three-position switch is provided for this purpose. Unidirectional response is sometimes obtained by the use of a large parabolic reflector mounted behind a conventional microphone. A parabolic reflector microphone also has higher sensitivity and is more nearly independent of room acoustics than is a conventional microphone.

4.2   MICROPHONE PLACEMENT

When a piano solo is to be broadcast, the optimum microphone placement must be determined experimentally. An omnidirectional microphone is suitable. As depicted in Fig. 4–2, the microphone is generally placed about ten feet from the piano. Depending upon the performer's manner of playing the bass and treble notes, the microphone may be moved nearer the bass strings, or nearer the treble strings. Note that the comparative loudness of the bass and treble notes also depends to some extent on the acoustics of the studio. A microphone is placed to pick up a piano duet as shown in Fig. 4–3. Volume balance is obtained by moving the microphone closer to one piano, and farther away from the other. Treble and bass equalization is obtained by rotating the position of either piano as may be required.

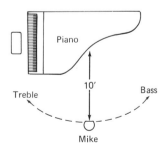

Fig. 4–2. Basic microphone placement for piano pickup.

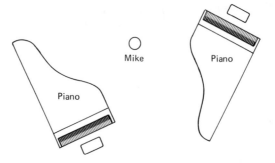

Fig. 4–3. Microphone placement for a piano duet.

Pickup from a vocalist with piano accompaniment is generally made as depicted in Fig. 4–4 with a bidirectional microphone placed between the vocalist and the piano. The microphone is typically spaced about eight feet from the piano and eight or ten feet from the vocalist. However, the relative spacings will vary depending on the comparative loudness of the two sound sources. Note that if the microphone is placed too near the sources, the dynamic range of the performance will become excessive and will tend to overload the audio system. Although overload can be corrected when required by monitoring and reducing the setting of the volume control at intervals, it is good practice to place the microphone so that minimum attention is needed by the monitoring operator.

A small orchestra is often arranged as shown in Fig. 4–5. The violins are typically located about four feet apart and the microphone is placed approximately four feet in front of the violins. This

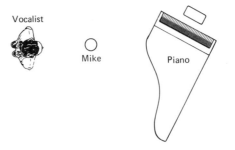

Fig. 4–4. Bidirectional microphone placement for vocalist with piano accompaniment.

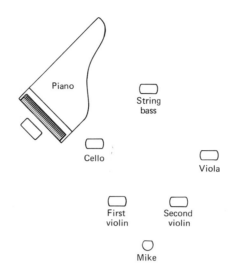

Fig. 4–5. Microphone placement for a small ensemble.

placement generally provides good tonal balance. However, in case
the violins are too loud, they may be moved back somewhat. Or,
the microphone may be raised considerably, so that it "looks down"
on the ensemble. Conversely, if the violins need to be emphasized,
they should be left in their original positions (Fig. 4–5) and the
microphone should be lowered near the floor. This effectively brings
the microphone nearer the violins with respect to the other instruments in the ensemble. A cardioid microphone is suitable, although
noise pickup and echoes are seldom a problem in this situation. In
other words, an omnidirectional microphone may be utilized.

Next, consider microphone placement requirements for an orchestra with soloist and chorus, as depicted in Fig. 4–6. It is advisable to use two bidirectional microphones to achieve better tonal
balance. One microphone responds chiefly to the chorus and soloist,
and the other microphone responds chiefly to the orchestra. In turn,
the outputs from the two microphones are mixed at levels chosen
by the program director. It is desirable to minimize the amount of
orchestral pickup by microphone 1 and the amount of choral and
soloist pickup by microphone 2 since there is a phase difference
between the microphone outputs from the two sources. This source
of audio waveform distortion is particularly severe when the microphones are oppositely polarized. Therefore, it is also good practice
to make certain that the microphones are polarized similarly.

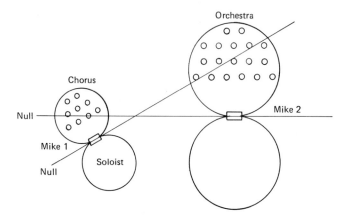

Fig. 4–6. Placement of bidirectional microphones for soloist, chorus, and orchestra.

Figure 4–7 exemplifies microphone placement for a large orchestra. It is preferable to use a single microphone, placed about as far back from the conductor as the depth of the orchestral group. The acoustics of the studio can be controlled to a considerable extent by means of movable drapes. It is desirable to have a com-

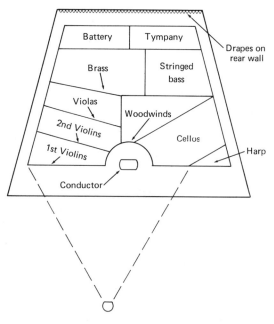

Fig. 4–7. Microphone placement for a large orchestra.

paratively "dead" studio, although some reverberation is needed. In other words, the studio should not approximate an anechoic chamber, or else the harmonics of the tones will sound weak and the performers will likely be uncomfortable. Although some reverberation is desirable, echoes are to be avoided. If these principles are observed, the pickup of a program from a large orchestra presents few or no problems. A cardioid microphone is basically appropriate, although an omnidirectional microphone will usually serve satisfactorily also.

4.3   VOLUME INDICATORS

As noted previously, a VU (volume unit) meter is employed as a volume indicator. Figure 4–8 illustrates a common type of VU meter and Fig. 4–9 depicts a standard volume-indicator circuit. A VU meter is basically a dB meter with a response characteristic that is best suited to monitoring complex audio waveforms. Zero VU indicates a reference level of 1 milliwatt of sine-wave power in 600 ohms. In other words, a VU scale is essentially a dBm scale (zero dBm denotes one milliwatt in 600 ohms). To extend the basic range of a VU meter, an external multiplier in the form of a T pad with 11 steps of 2 dB each is utilized, as depicted in Fig. 4–9. The VU meter works out of a 600-ohm load, as 600-ohm lines are standardized in broadcast audio systems.

Fig. 4–8.   A widely used type of VU meter. (Courtesy of Simpson Electric Co.)

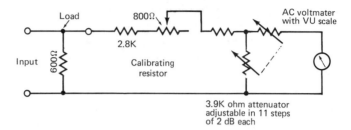

Fig. 4–9. A standard volume-indicator circuit.

4.4   SPEECH-INPUT AMPLIFIERS

A speech-input amplifier is also called a preamplifier, microphone, line, or program amplifier. Approximately 75 dB gain is provided, with a frequency response from 30 to 15,000 Hz. In other words, a speech-input amplifier is basically a high-fidelity unit. Thus, the design of the amplifier is essentially the same as previously described for high-fidelity preamplifiers.

4.5   VOLUME-COMPRESSION AMPLIFIERS

Volume-compression amplifiers, also called limiting amplifiers, are used to automatically reduce the gain of the audio system when the program peaks are unusually high, thereby avoiding overmodulation and distortion (*sideband splatter*). Figure 4–10 shows the plan of a volume-compression amplifier. It has a maximum available gain of 54 dB, with a frequency response that is flat within 1 dB from 30 to 10,000 Hz. Compressor action is basically AVC control, with a certain threshold and a certain attack time, followed by a certain recovery time. In other words, the bias voltage applied to the rectifier determines the threshold of compressor action. If a program peak exceeds this threshold value, a reduction in gain occurs in 1 millisecond. In turn, the compressor control bias leaks off slowly to ground in approximately 7 seconds. Or, when faster recovery is desired, the switch to the slow filter may be opened, so that the fast filter only is in the circuit.

4.6   SPEECH CLIPPERS

When only speech is being transmitted, it is usual to clip the peaks that would cause overmodulation, without causing serious distor-

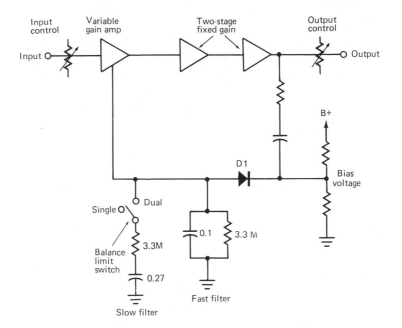

Fig. 4–10. Plan of a volume-compressor amplifier.

tion of the information content. In this application, a speech clipper permits operation at a higher average-signal level than if a compression (limiting) amplifier were utilized. Note that a clipper is virtually instantaneous in attack, whereas a typical compressor requires a millisecond of attack time. When a speech clipper is used, it is standard practice to attenuate the audio signal above 4 kHz, with 25 dB attenuation at 10 kHz. Thereby, the prominent harmonics that are generated in the clipping process are attenuated or removed. Approximately 20 dB of peak clipping is possible before distortion becomes serious. The corresponding increase in average audio power level is approximately 12 dB. A basic type of speech clipper is depicted in Fig. 4–11. The clipping level is changeable by adjustment of R1 and a compensating adjustment of R2. In radio-broadcast technology, a combination of compression action and clipper action is often employed.

4.7   EQUALIZERS

Equalizers are used in broadcast audio systems to compensate for frequency distortion in cables, lines, or loads. The simplest type of

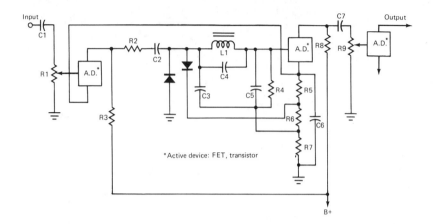

Fig. 4–11. A basic type of speech clipper.

equalizer is a differentiating or integrating circuit inserted at the receiving end of the cable to obtain low-frequency attenuation or high-frequency attenuation. Most cables impose a progressive attenuation of the higher audio frequencies, as exemplified in Fig. 4–12. In turn, an equalizer is employed which imposes a progressive attenuation of the lower audio frequencies. Although the system frequency response cannot be made perfectly uniform, a great improvement can be obtained by selecting a suitable equalizer circuit with correct component values. One of the limitations of an

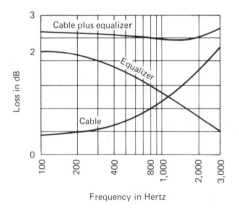

Fig. 4–12. Cable, equalizer, and system frequency responses.

equalizer is its inability to compensate for load changes, such as may occur during switching procedures. However, equalizers may also be switched in or out of the load circuit. An equalizer has a definite insertion loss, but this loss may be compensated by additional amplification. Figure 4–13 shows the configuration of a typical equalizer.

4.8   REPEATING COILS

Repeating coils are basically audio transformers. They are used for impedance transformation, for DC isolation, and for various other functions. However, in broadcast audio systems, they are utilized chiefly for impedance transformation. The impedance ratio of a transformer is equal to the square of its turns ratio. For example, if a transformer has twice as many turns on its secondary as on its primary, the impedance ratio from primary to secondary is 1-to-4. Figure 4–14 shows how repeating coils (audio transformers) are used to step up a low-source impedance to a higher-line impedance and, conversely, to step down a higher-line impedance to a lower-load impedance. Of course, a high-source impedance may be matched to a lower-line impedance and thence to a higher-load impedance when required. Other basic applications for repeating coils are explained in a following chapter.

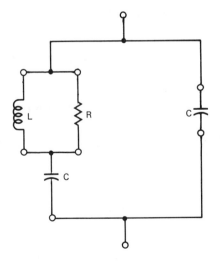

Fig. 4–13. Typical equalizer configuration.

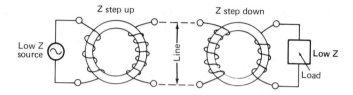

Fig. 4–14.   Source-to-line-to-load matching system.

4.9   ATTENUATORS

Various types of attenuators are used in broadcast audio systems. Continuous attenuators generally employ a *T* or *H* configuration, as explained in the first chapter. The *H* configuration is utilized when the audio circuits must be balanced to ground for the purpose of minimizing crosstalk and noise that may be present in unbalanced circuits. The chief advantage of a *T* or *H* attenuator over simpler configurations is maintenance of constant input and output impedance at any setting. Although most *T* and *H* attenuators have the same value of input and output impedance, this is not necessarily the case. For example, this type of attenuator can be designed for an input impedance of 600 ohms and an output impedance of 8 ohms.

Step-type attenuators are also widely used in broadcast audio systems. The multiple-*T* step attenuator shown in Fig 4–15 has the advantage of zero insertion loss in its first position. This feature is sometimes an advantage in audio circuitry. The input and output impedances for this type of attenuator are equal and are the same on any step. When large ranges of attenuation from a high-level to a very low-level output are required, the ladder attenuator depicted in Fig. 4–16 is preferred. Although the input resistance is constant on any step, the output impedance varies somewhat. Both multiple-*T* and ladder attenuators may be designed for any desired number of steps. In broadcast applications, 2-dB steps are often employed with a total range of 20 dB (11 steps).

4.10   CONSTANT-IMPEDANCE FADERS AND MIXERS

T-type attenuators are connected in series or parallel to provide attenuation and mixing. When the output from one attenuator is

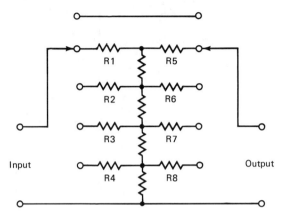

Fig. 4–15.   Multiple-step *T* attenuator with a zero-insertion-loss position.

increased as the output from another attenuator is decreased, the arrangement operates as a fader. Figure 4–17 shows typical series and parallel configurations. They generally have 600-ohm input and output impedances, although 500, 200, or 150 ohm arrangements are occasionally used. Note that the variable or adjustable attenuators are followed by a fixed resistive pad. This pad provides an output impedance which is equal to the input impedance of the configuration. Pads utilized for impedance matching are often called *taper pads*.

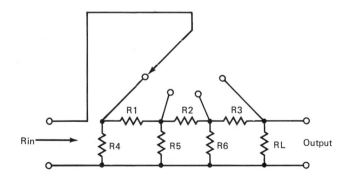

Fig. 4–16.   Example of a ladder attenuator.

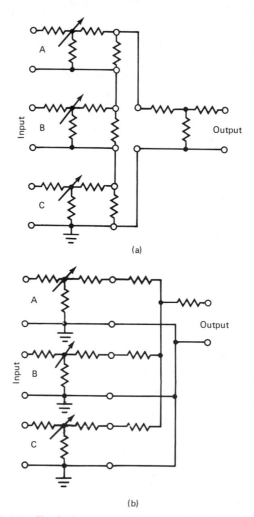

Fig. 4–17. Typical constant-impedance fader-mixer configurations. (a) Series connection of *T* attenuators. (b) Parallel connection of *T* attenuators.

Figure 4–18 shows a plan for a single program channel. Three studios are provided for, with a preamplifier and fader for each. Signals from an individual studio can be switched off or can be mixed with signals from the other studios. A studio master fader and a nemo master fader are also provided. The outputs from these faders are fed into a common output line along with the output from a telephone company line. Time signals are introduced

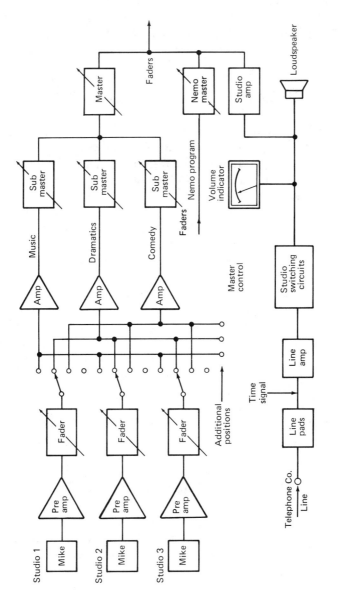

Fig. 4–18. Plan of a single program channel.

into the telephone line output, prior to the studio switching facilities. Separate monitoring equipment is employed for the telephone signals, following the studio switching facilities. In turn, program-channel monitoring equipment is utilized in the common output line (not shown in Fig. 4–18).

4.11   PROGRAM RECORDING FACILITIES

Broadcast audio systems require program recording facilities for reference recordings. These are essentially a log of the program material that is transmitted. Recording facilities are also required for delayed broadcasts of special-events programs. In addition, recordings may be produced for use at small broadcasting stations that lack wire-line facilities. Recordings are also required for auditions and/or rehearsals. Recording facilities are generally used for sound effects inserted into various program sequences. Tape and disc recordings are employed, using equipment similar to the units discussed in the first chapter. Discs are cut in a machine device similar to a lathe with a lead screw. A horizontal turntable supports the wax blank that is to be cut. A sapphire cutting stylus is often used, driven laterally by an electrodynamic transducer. A playback transducer is also provided for playing back the disc immediately after cutting for rehearsal operations and tests. Since playback of the wax disc generally deteriorates the grooves so that it is unsuitable for electroplating and pressing, it is customary to make two wax recordings at the same time. In turn, one may be used for immediate playback and the other can be electroplated for pressing many duplicates.

4.12   FIELD AUDIO FACILITIES

Basic field audio facilities are depicted in Fig. 4–1. Either a telephone line or cable may be utilized between the remote pickup site and the studio, or a portable radio transmitter may be employed to link a mobile unit with a radio receiver at the studio. Figure 4–19 illustrates a broadcast mobile unit, with its antenna. Microphones utilized in field audio facilities are weatherproof and have comparatively rugged construction. Highly directional and sensitive microphones such as shown in Fig. 4–20 are employed when it is necessary to pick up sounds from comparatively distant sources.

Fig. 4–19.   Typical interior of a broadcast mobile unit. (Courtesy of RCA)

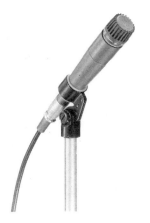

Fig. 4–20.   Sensitive directional microphone used in field audio facilities. (Courtesy of Electro-Voice, Inc.)

Fig. 4–21. Portable transmitter/receiver used by announcer in field audio activities. (Courtesy of Realistic)

Portable transmitter/receivers, such as illustrated in Fig. 4–21,
are also used in the field to link the announcer to a mobile unit or
directly to a radio receiver at the studio. If the announcer can
remain within a reasonable distance from the mobile unit, cables
are used to connect with the microphone and the cue-monitoring
headset.

4.13   TROUBLESHOOTING BROADCAST
AUDIO EQUIPMENT

Troubleshooting broadcast audio equipment involves the same approaches and procedures discussed in previous chapters, with one
important addition. Since down-time must be strictly minimized,
or eliminated if possible, preventive maintenance is of great im-
portance. In other words, microphones, cables, and all key units
should be checked out before *going on the air* each day. Although
strict maintenance schedules will turn up many defects in preair
checks, trouble symptoms will nevertheless occur at times while a
program is being aired. In turn, an emergency situation arises, unless a spare microphone, cable, preamplifier, program amplifier,
monitor amplifier, power supply, or switching facility is available, as

the occasion demands. Since it is impractical and uneconomical to duplicate a complete broadcast audio system, the burden of an emergency will fall at times upon the operator. His duty is to pinpoint the defective unit or component as rapidly as possible and get the station *back on the air.*

It is essential that the operator be completely familiar with the equipment charged to his responsibility. In other words, he needs to have block diagrams and signal flowcharts memorized. The function of each unit and its relation to the functions of other units must be thoroughly understood. For example, the operator in charge of a control console, such as depicted in Fig. 4–22, should be able to draw the complete diagram with total accuracy from memory. There is no time for consulting instruction books and reading troubleshooting directions during emergency situations. A competent operator has invested considerable time and study in anticipating all types of component failure, the trouble symptoms that result, and the tests or measurements that should be made in each case. Some operators who are otherwise qualified tend to panic under severe emergency conditions with the result that the station remains *off the air* until a factory technician is found.

4.14   TROUBLESHOOTING PA SYSTEMS

Troubleshooting public-address systems involves many of the approaches and procedures described in the previous chapters. Common trouble symptoms include complete lack of output, failure of part of the system, weak output, distorted reproduction, and intermittent operation. A *dead* system is most likely to result from an amplifier defect, although a cable break is sometimes responsible. In a simple PA system employing a single microphone, the possibility of microphone trouble, or cord defect should be checked at the outset. If a spare microphone is available, it can be plugged into the amplifier for testing. When a substitution test cannot be readily made, the microphone output can be checked with an audio voltmeter. In case that the trouble is localized to the amplifier, the same general troubleshooting procedures are followed as described in the first chapter.

Loudspeakers occasionally become defective, although this is not a highly probable occurrence. Since most PA systems utilize several or many speakers, there is little difficulty in identifying a faulty unit. The only exception occurs where a short-circuit occurs

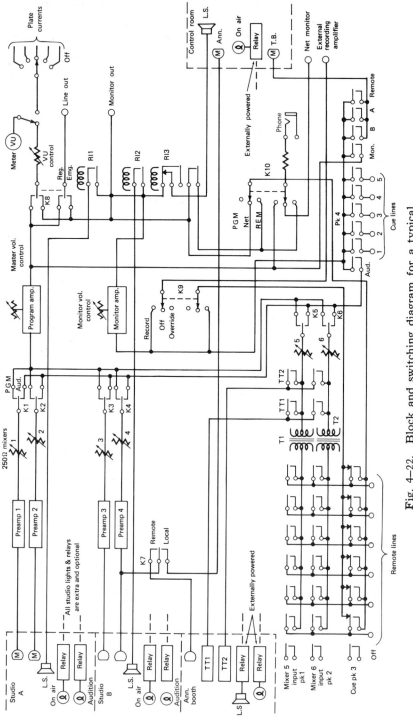

Fig. 4–22. Block and switching diagram for a typical control console. (Courtesy of RCA)

in the speaker, resulting in zero audio voltage across the speaker lines. Note that a milliohmmeter is the most useful instrument for localizing a short-circuit. Since the speaker lines have a small value of conductor resistance, this resistance can be checked with a milliohmmeter. By measuring the resistance values at the various output terminals, the short-circuited termination is found by noting the pair of terminals with the lowest resistance value. The result is confirmed by disconnecting the defective speaker from the line, which should result in the remaining speakers resuming operation.

Failure of part of a PA system is most likely to be caused by a cable defect or by a faulty speaker. If cables are run in conduit, there is little likelihood of trouble due to conductor breaks. However, the majority of PA systems employ exposed cable, which may be subject to mechanical damage by careless workmen or by vandals. In a 70.7-volt or 25-volt system, a break in the line can be localized without undue difficulty because all the speakers past the open-circuited point are *dead*. When multi-conductor cables need to be checked out, they can be tested by pairs for continuity with an ohmmeter, as explained in the previous chapter. In turn, the open-circuited conductor can be identified. It follows from previous discussion that there is a certain amount of overlap among PA, paging, and intercom systems. Thus, troubleshooting techniques tend to be somewhat similar for these systems.

Distorted reproduction is most likely to be caused by amplifier defects, although microphone faults or speaker faults are occasionally responsible. Amplifiers are serviced as explained previously. A substitution test is advisable when a microphone is suspected of producing distorted output. Speaker distortion is often accompanied by rattles, buzzing, and/or popping sounds. Reconing is often required, and voice coils may become damaged and require replacement. If a speaker is badly deteriorated, it is more economical to replace the unit instead of attempting to repair it. Intermittent operation is by far the most challenging trouble symptom for a technician. It can occur at any point in a PA system. As already noted, troubleshooting is practical only during the inoperative interval of the intermittent condition. Signal-tracing procedures are basic in localizing the trouble areas and the pinpointing of defective components is accomplished as explained in the preceding chapters.

Some PA systems employ an AVC circuit as depicted in Fig. 4–23. This arrangement maintains a more nearly uniform volume level as the operator moves closer to or farther way from the micro-

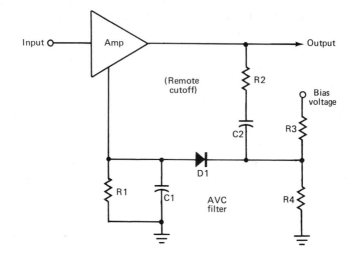

Fig. 4–23. AVC arrangement for a PA system.

phone. In case of poor or no control of volume, capacitors C1 and C2 should be checked for leakage or open circuit. If the rectifier diode develops a poor front-to-back ratio or some other defect, the AVC action will be impaired. Although AVC trouble symptoms can be caused by off-value resistors, this is a much less likely type of defect. Note that correct bias voltage must be applied to the rectifier diode to establish the optimum threshold of AVC action.

5

COMMERCIAL
TELEPHONE SYSTEMS

5.1   GENERAL CONSIDERATIONS

Telephony is the science of electrically transmitting speech between two or more points. Commercial systems employ carbon
transmitters (microphones) and watchcase-type receivers, as depicted in Fig. 5–1. These units are mounted in opposite ends of a
handset. They are used in a basic telephone circuit as shown in
Fig. 5–2. Induction coils are used for two reasons, both of which
contribute to operating efficiency. First, the induction coil restricts
the flow of DC current to the microphone or transmitter, thereby
avoiding the I^2R losses that would occur if the DC current flowed
through the telephone line. Second, the induction coil transforms
the comparatively low impedance of the carbon transmitter to the
higher impedance of the line, thereby providing maximum power
transfer of the audio signal. Ordinary receivers and transmitters
have a resistance of approximately 75 ohms each.

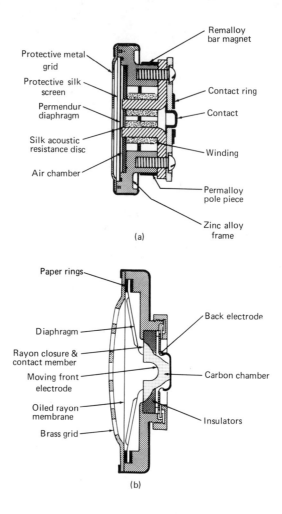

(a)

(b)

Fig. 5–1. Construction of commercial telephone receiver and transmitter (microphone). (a) Receiver. (b) Transmitter.

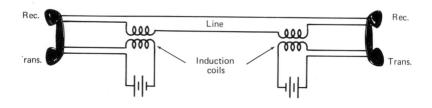

Fig. 5–2. Plan of a basic telephone circuit.

Fig. 5–3.   Appearance of a telephone induction coil.

A typical telephone induction coil is shown in Fig. 5–3. It consists of a core made of annealed soft-iron wires and wound with a primary of comparatively large wire and a secondary winding of many turns of small wire. For example, the primary may have 300 turns, and the secondary 1,100 turns. The primary resistance is about 1 ohm, and the secondary resistance is approximately 20 ohms. The impedance ratio in this example is 13.3–to–1. An induction coil is designed to have a frequency response of approximately 300 to 3,400 Hz. This frequency range is adequate for acceptable articulation. Note that conversational speech actually has a frequency range of 62 to 8,000 Hz. However, it would be uneconomical to transmit the complete speech-frequency range.

5.2   COMMON-BATTERY CONFIGURATION

Induction coils (generally called *repeating coils*) are widely used in the common-battery system depicted in Fig. 5–4. Both transmitters are energized by DC current from the line. The audio-

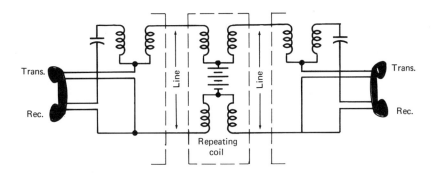

Fig. 5–4.   Basic common-battery system.

current flow in one line is repeated into the other line with little energy loss. Note that blocking capacitors are inserted in series with the receivers to prevent shunting DC current from the line. The chief disadvantage of the basic arrangement shown in Fig. 5–4 is the fact that when the transmitter of a handset is spoken into, the receiver in the same handset reproduces the audio signal quite loudly. This is distracting to the user, and repeating coils are generally elaborated to minimize the reproduction of *sidetone*. Figure 5–5 shows a simple *antisidetone* configuration. This arrangement permits the receiver to respond only to audio signals originating from the transmitter at the other end of the line. The theory of operation of an antisidetone circuit is as follows.

Refer to Fig. 5–5. The antisidetone circuit employs an induction coil with two secondary windings. Winding $S2$ has a high resistance compared to winding $S1$. In turn, $S2$ does not have an appreciable shunting effect across the receiver. When the transmitter is spoken into, there are two parallel paths for audio current flow. One path is directly into the line through the primary of the induction coil; the other path is through the capacitor and $S1$, and either through $S2$ or the receiver. However, the induction coil is polarized and wound so that current flowing in P and $S1$ will induce a voltage in $S2$ which tends to cancel the former flow of current flowing through the receiver. In turn, the amount of sidetone is minimized. Complete cancellation can be obtained at only one frequency so that there is residual sidetone present, although it is quite weak. Note that the ringer bell windings are connected in series with the capacitor across the line.

5.3   COMMON-BATTERY EXCHANGE

The basic arrangement of a common-battery exchange with two stations is shown in Fig. 5–6. Note that the telephone circuit at

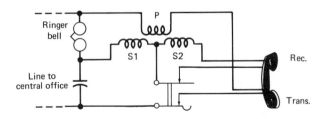

Fig. 5–5. Anti-sidetone circuit.

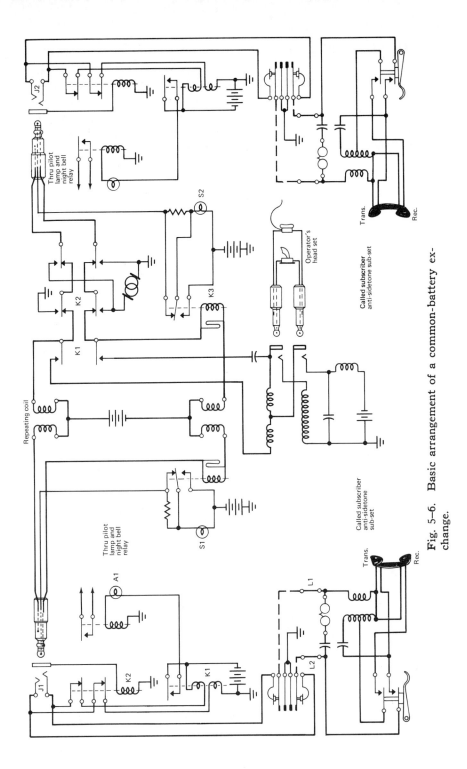

Fig. 5-6. Basic arrangement of a common-battery exchange.

133

each station is normally open when the handset is on the switch (hook). However, the ringer is bridged across the line in series with a capacitor. The operator calls the subscriber by a ringing signal, and the subscriber calls the operator by lifting the handset and thereby closing the switch. Contacts $C1$ and $C2$ are made at the switch. $C1$ closes the line through the transmitter in series with the primary of the induction coil. In turn, current flows from the central-office battery B through one-half of the line relay winding $R1$, over one side of the line $L1$, through the primary of the induction coil and the transmitter, back to the central office over the other half of the relay winding $R1$, to the other side of the central-office battery or to ground (the grounded end of the battery).

The foregoing circuit closure energizes the line relay $R1$ which connects the central-office battery to a small answer lamp $A1$ in the face of the switchboard in front of the operator. Lighting of this lamp indicates to the operator that this particular line is calling. In turn, the operator answers the call by inserting plug $P1$ into the jack associated with the lighted lamp and to which the line of the calling party is connected. Next, a third battery connection to the sleeve of the plug closes a circuit through the winding of a second relay $R2$, called the *cut-off* relay, which disconnects the line relay from the circuit. This sequence turns off the answering (or line) lamp $A1$. The operator connects her telephone set to the cord circuit by means of the listening key $K1$. When the operator speaks into her transmitter, the audio currents flow through the two heavy conductors of the cord circuit through the windings of the repeating coil, which, by transformer action, induces current into the other windings of the coil. This audio current flows back over the calling subscriber's line and induces a current in the secondary of the induction coil, which flows through the telephone receiver.

Note that the audio signals from the operator flow from the central-office cord circuit to the subscriber's receiver and also note that there is a direct current supplied by the central-office battery through two of the four windings of the repeating coil in the cord circuit, over the line, and through the subscriber's transmitter. This current is equivalent to the transmitter current supplied by the local battery in the arrangement of Fig. 5–2. It permits generation of audio currents by variation of the transmitter carbon resistance, which, by means of the repeating coil windings at the central office, induces an audio current across to the opposite side of the cord circuit.

On being informed of the number of the party called, the operator inserts plug *P*2 into jack *J*2 which turns on lamp *S*2 by closing the circuit from the central-office battery through the sleeve connection and the cut-off relay winding. This lamp indicates that the handset is on the switch (hook) at the called party's station and that she must attempt to obtain the called party's attention by ringing, which is accomplished by operating the ringing key *K*2. When the called party answers, current flowing from the central-office battery through the windings of the repeating coil and through the supervisory relay *R*3 serves to operate this relay. In turn, lamp *S*2 is short-circuited and is turned off, indicating to the operator that the party has answered. Simultaneously, a resistance is inserted into the battery circuit to limit the current flowing through the cut-off relay. The standard ringing frequency of 20 Hz is functionally a subaudio frequency.

After both parties have finished conversing and have replaced their handsets, this supervisory relay, as well as a similar relay on the other side of the cord circuit, becomes deenergized. Since the short-circuit is then removed from the lamps, they are turned on. Lighting of the lamps indicates to the operator that both parties are finished conversing, and that both cords are to be taken down. When the operator pulls down both cords, the sleeve circuit of the cord is opened and the lamps are extinguished. Thus, the operator depends upon glowing lamps for each operation except that of connecting the calling cord to the jack of the called station. In all common-battery operations, a glowing lamp means *attention*. In other words, a glowing lamp in the face of a switchboard indicates *line to be answered*. A glowing lamp at a cord indicates *continue ringing on the corresponding cord*. Two lamps glowing indicates *disconnect both cords as both parties have replaced their handsets*. However, a flashing lamp indicates that one party is not replacing his handset, but wishes to place another call, or desires the operator to answer.

5.4   PHANTOM CIRCUITS

To obtain additional communication channels in a telephone line system, phantom circuits are employed, as shown in Fig. 5–7. This arrangement provides an additional telephone channel for each four wires, or an increase of 50 percent in communication capacity. Note that the four repeating coils and telephone lines are connected into a basic bridge circuit. The conventional channels are called phys-

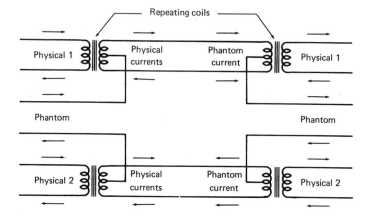

Fig. 5–7. Two physical circuits with one phantom circuit.

ical circuits and the added channel is called a *phantom circuit.* If an audio signal is applied to the input of the physical No. 1 circuit, the signal flows to the output of the physical No. 1 circuit. On the other hand, the signal does not appear at the output of the physical No. 2 circuit, nor does it appear at the output of the phantom circuit. The same relations hold for the No. 2 channel. If an audio signal is applied to the input of the phantom channel, the signal flows to the output of the phantom channel, but this signal does not appear at the output of the physical No. 1 channel nor at the output of the physical No. 2 channel.

5.5 LOADING OF TELEPHONE LINES

An ideal telephone line would have no losses. Practical lines have losses due to conductor resistance, leakage between conductors, and conductor impedance. Impedance results from the capacitance between the conductors and to a lesser extent from the self-inductance of the conductors. Resistance is determined entirely by the kind of metal used and the diameter of the conductors; of course, a long line consisting of a given type of conductor will have more resistance than a short line with the same type of conductor. Leakage between conductors can be minimized by good manufacturing and installation practices, although it cannot be entirely eliminated. Capacitance by itself tends to bypass the higher audio frequencies.

Inductance by itself tends to oppose the passage of the higher audio frequencies. Note, however, that if inductance and capacitance are ideally proportioned in a line, their effects cancel out so that there is no attenuation of the higher audio frequencies in the ideal line.

In an ideal line, the inductance and capacitance are distributed, or, there is no variation of the inductance and capacitance values from one point of the line to the next. In a practical line, the inductance and capacitance are also distributed. However, unlike the ideal line, a practical line also has conductor resistance and more or less leakage resistance. The effect of the resistance is to cause an effective disproportion of the amount of inductance and capacitance in the line. If inductors (coils) of suitable value are inserted in series with the line at fairly frequent intervals, a better balance of inductance and capacitance can be obtained. In turn, the high-frequency attenuation of the telephone line is reduced. This technique of improving high-frequency response is called *loading*.

Figure 5–8 shows the basic method of loading a telephone line. A loading coil consists of two windings on an iron core. The windings are polarized so that they magnetize the core in the same di-

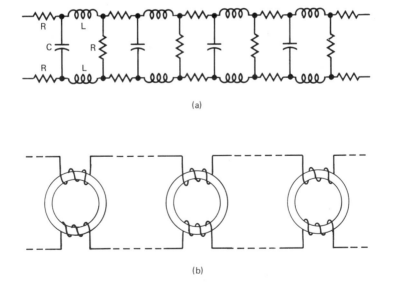

(a)

(b)

Fig. 5–8. Principle of loading a telephone line. (a) Approximate equivalent circuit of a line. (b) Basic loading-coil arrangement.

rection in order that the mutual induction of the coils adds to the self-induction of each. In this way, smaller coils may be employed. Low resistance wire is used and toroidal construction is utilized to make the coils unresponsive to external fields. The two coils are very closely matched so as to maintain good line balance. If loading coils are connected into the line at frequent intervals, improved high-frequency response is obtained, compared to a line with coils inserted at widely distant intervals. In practice, three types of loading are employed, called *heavy*, *medium*, and *light* loading.

Heavy loading denotes the use of 0.22 henry of inductance per mile, with a spacing of 1.14 miles between consecutive coils. Medium loading denotes 0.105 henry per mile at intervals of 1.66 miles. Light loading denotes 0.061 henry per mile at intervals of 2.2 miles. Heavy loading provides great advantage over no loading; a typical unloaded telephone cable is usable for only 30 miles before amplification of the signal is required. On the other hand, heavy loading enables the same cable to be installed 150 miles before amplification is required. Medium loading permits 90-mile cable runs. Heavy loading is comparatively costly both in terms of material and labor of installation. Figure 5–9 depicts the connections employed when a phantom loading coil is connected into a pair of lines.

5.6 TELEPHONE REPEATERS

When the audio signal becomes greatly attenuated, the signal-to-noise ratio becomes poor. If the signal becomes excessively attenuated, it will be masked by noise. Therefore, the signal-to-noise ratio must not be permitted to fall below a tolerable level. This requires the use of amplifiers at approximately 50-mile intervals along telephone cable circuits. A special type of amplifier called a repeater is utilized. For example, the 22-type repeater is in general use. The "22" term denotes "two-element, two-way". This means that there are two distinct one-way amplifiers employed in the repeater and that it is installed in an ordinary two-way telephone circuit. A repeater has the following features:

1. A repeater employs one-way amplifier circuits designed to provide the necessary amplification and is equipped with regulating devices for adjusting the gain to meet a range of operating conditions.

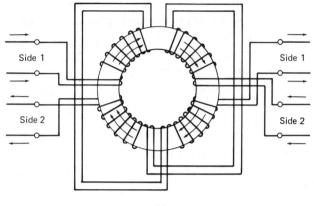

(a)

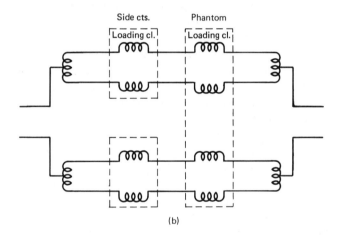

(b)

Fig. 5–9. Connection of a phantom loading coil into a pair of telephone lines. (a) Windings of a phantom loading coil. (b) Connections for loading both the side circuits and the phantom circuit.

2. Special transformers called hybrid coils are used to provide two-way transmission with one-way amplifiers.

3. Proper network balancing equipment is utilized for closely approximating the impedances of each line circuit and its associated apparatus over the audio range in order to maintain the balance required by the hybrid coil for proper operation. In other words, audio frequencies from the output of

one amplifier must not enter the input of the other amplifier because the speech quality would be impaired or the system might even *sing* (oscillate).

4. Filters are employed for eliminating any energy not within the audio-frequency range.

5. Miscellaneous apparatus and circuit features are also used to adapt a telephone repeater circuit to standard operating practices.

Figure 5–10 depicts a simplified diagram for a 22-type repeater with the features noted above. Operation of the repeater is as follows:

We will assume that the subscriber at the east end of the line is talking and that the substantially attenuated audio signal from his station reaches the telephone repeater circuit at the hybrid coil on the east side of the repeater. Half of this entry is transmitted from terminals $3T$ and $8T$ of the coil through the pad, which further attenuates the audio current in accordance with the relative values of X and Y to the east-west potentiometer bridged across the input circuit. Note that the adjustable element of this potentiometer comprises a double slide-wire with a 200-ohm resistor inserted between its halves. A shunt resistance of 1212 ohms, grounded at its mid-point, is inserted ahead of the potentiometer, which gives the circuit an impedance of approximately 300 ohms as seen from the hybrid-coil side.

The internal balance of the repeater in Fig. 5–10 is improved by grounding the shunt resistor at its midpoint. The double slidewire contacts, which are used in part to control the gain of the amplifier, *pick off* from half to the full voltage drop across the potentiometer and apply it to the low-impedance winding of the input transformer. In case that half of this voltage drop is greater than is required to provide desired overall gain, a pad having a greater loss may be inserted in the input circuit. Thus, the voltage that has been picked off is stepped up approximately 30 times by the transformer and is applied to the control device where it serves to control the current flow in the output circuit. The output-circuit energy, which has the same waveform as the input-circuit energy but a much greater amplitude, then passes through a filter to terminals 2–5 of the west output transformer (hybrid coil).

Figure 5–11 shows the basic hybrid-coil arrangement. It is functionally a form of a bridge circuit. The audio voltage source is represented by E with an internal resistance R. z_1 represents the

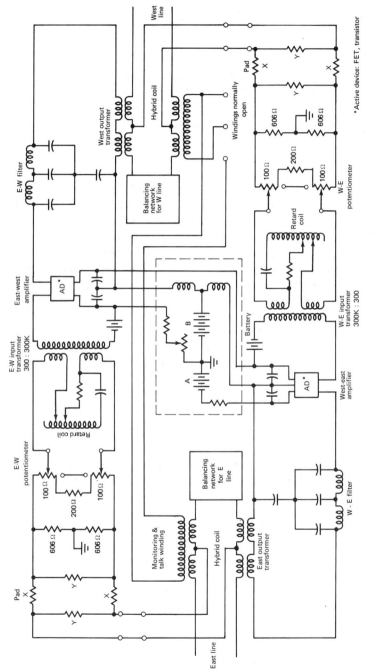

Fig. 5–10. Configuration of the 22-type telephone repeater.

141

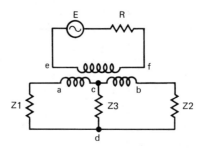

Fig. 5–11. Basic hybrid-coil arrangement.

input impedance of the line, z_2 that of the balancing network, and z_3 that of the input circuit of the amplifier. To prevent oscillation (singing), no current should flow in z_3 due to the voltage E. This will be the case if ac and cb are identical windings, and z_1 is equal to z_2. The winding ef has a turns ratio with respect to the other windings that matches impedances from input to output, thereby providing maximum audio power transfer. Since the input impedance at terminals a–d should match the line impedance, the value of z_3 is chosen accordingly. The load on the secondary windings is equal to $2z_1$. Thus, the amplifier must produce twice the gain that is required for actual transmission because one-half of the output power is lost in the balancing network.

If this network balances the line exactly, no part of the energy reaches the west-east amplifying circuit. On the other hand, if the balance is not perfect, a part of the energy will cross the hybrid coil, will be amplified, and then will be returned to the east line. Also, if there is some unbalance in the east-line network, a part of this returned energy will cross the east transformer and return in amplified form to the west end of the circuit. In case these unbalances are sufficiently large, the amplifier will start to sing (oscillate), and will become inoperative. Since the repeater configuration is completely symmetrical, its operation for transmission in the opposite direction may be followed in the same manner as above described.

5.7   DIAL SWITCHING ARRANGEMENTS

A dial switching system is an electro-mechanical arrangement that operates without the services of a switchboard operator. Its basic function, however, is the same as explained previously. When the dial is operated, it makes and breaks a DC circuit, thereby generating pulses. For example, if the number 8 is dialed, the mechanism generates a series of eight pulses. In turn, these pulses operate switching equipment at the central office. Some dial central-office switching equipment functions on a step-by-step plan. In other words, each digit that is dialed causes a movement of a switch at the central office and carries the connection one step forward to the called subscriber's line. This basic principle is diagrammed in Fig. 5–12. Two steps are shown, and there are five contacts on the switches for each step. If the switches move in sequence through the same number of contacts as the number of pulses arriving from the dial, any one of 25 separate telephones can be reached by dialing two successive digits between 1 and 5.

In the standard telephone dial system, there are ten digits and the system can process as many as seven digits to call a desired subscriber. This permits any one of ten-million subscribers to be called in seven steps of the dial. The chief disadvantage of this simple arrangement is that a tremendous number of switches and interconnections would be required to permit one telephone to call any of the other subscribers. Therefore, the basic system must be

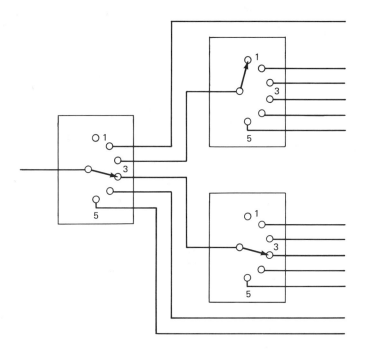

Fig. 5–12. Step-by-step switching configuration.

elaborated to meet practical requirements. First, the necessity is eliminated for having a separate selector switch for each subscriber's line. Instead, line-finder switches are employed for groups of subscriber lines. Each line-finder is connected in multiple to the *bank* terminals of the switches. In this step-by-step arrangement, the line-finder switch appears as depicted in Fig. 5–13; it can move vertically through ten steps, and ten steps horizontally for each vertical step. If a subscriber picks up his handset, a relay on the line causes an idle line-finder switch to start operation and to hunt for the terminal to which the subscriber's line is connected.

A simple dial system employing these principles is shown in Fig. 5–14. The line-finder switch connects the calling subscriber's line to a trunk-selector switch installed in the same central office. The step-by-step system has the same design as the line-finder switch that was shown in Fig. 5–13. When the caller lifts his handset, a dial tone is automatically sent back to the caller. As he starts to dial, the trunk-selector switch in Office *A* moves to a terminal at which a trunk corresponding to the first digit dialed is connected. In turn, this trunk may run to a distant office or to terminating equipment in the same office. In this example, the first switch at Office *B* selects a group of connecting lines to be controlled by the

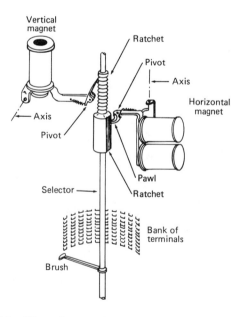

Fig. 5–13.   Plan of a step-by-step line-finder switch.

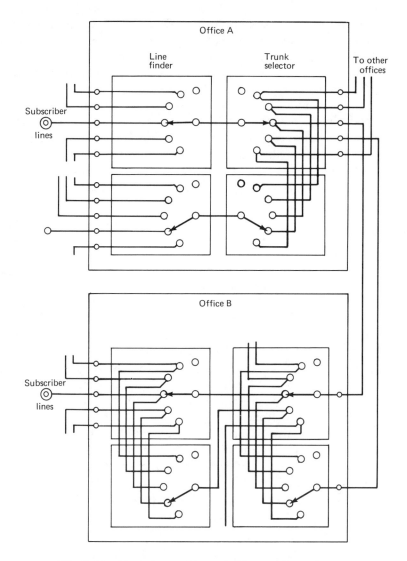

Fig. 5–14. Basic inter-office switching configuration.

second digit dialed and also moves the call on to a final selector switch which in turn will be operated by the third number dialed.

Notice in Fig. 5–14 that there are two trunks (interoffice connecting cables) between Office *A* and Office *B*, each of which connects to selector switches in the two offices. In turn, a switch finding a called trunk busy will automatically move on and hunt for another trunk to the same called office. The number of trunks required depends on the maximum number of subscribers who are likely to place calls during the busiest hour. A maximum of ten trunks can be accommodated by the line-finder switching system. If more than ten trunks are required, a faster type of dial switching system must be employed. These are elaborated forms of electro-mechanical selector switches that in turn have certain limitations on speed of operation. Therefore, sophisticated electronic switching systems are making steady inroads into telephone central offices since an electronic switching installation is more than six times as fast as the most elaborate electro-mechanical installation.

5.8 ELECTRONIC SWITCHING

Transistors are generally used as electronic switching devices. In addition to high-speed operation, transistors also provide amplification, can reverse the polarity of the output signal if desired, and can permit approximate matching of input and output impedances for optimum power transfer. When operated in the switching mode, a transistor is either open-circuited (cut off), or it is short-circuited (saturated), depending on the base-emitter bias voltage that is applied to the transistor. A typical transistor switching circuit is depicted in Fig. 5–15. Switch S1 controls the polarity and value of base current from battery V_{B1} or V_{B2}. Resistors R_{B1} and R_{B2} are current-limiting resistors. The emitter-base and collector-base diode and switch equivalent circuits representing the *off* and *on* (DC) conditions of the transistor switching circuit are shown in (*b*) and (*c*) of Fig. 5–15.

Note in Fig. 5–15(*b*) that the off condition of the circuit corresponds to reverse biasing of both diodes. In other words, V_{B2} is polarized in a direction that reverse biases CR_E, and V_{CC} is polarized in a direction that reverse biases CR_C. Therefore, no current flows in the circuit. On the other hand, in the on condition of Fig. 5–15(b), a forward bias is applied to CR_E by V_{B1}. In turn, current flows through CR_E. Moreover, the polarity of V_{B1} also has the effect of applying a forward bias to CR_C, and current flows through CR_C. This current flow produces a voltage drop across R_L and an output voltage is thereby obtained. Figure 5–15(c) shows the equivalent mechanical switching circuit.

Figure 5–16 illustrates how a pulse is produced from step-voltage waveforms. Switching operations always produce pulses.

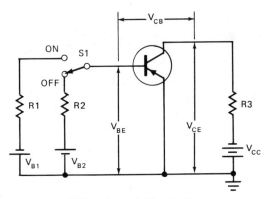

(a) Transistor switching circuit

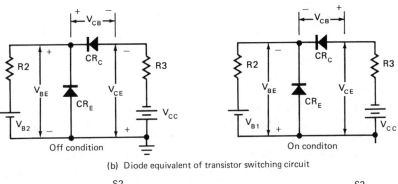

(b) Diode equivalent of transistor switching circuit

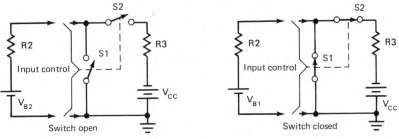

(c) Switch equivalent of transistor switching circuit

Fig. 5–15. Basic transistor switching circuit and equivalent circuits.

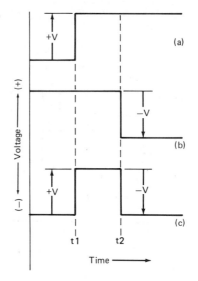

Fig. 5–16. Step-voltage waveforms, showing the formation of a pulse.

Note that if a pulse voltage is passed through a transistor switching circuit, the pulse is reproduced in amplified form in the output circuit. If the switching transistor is operated in the common-emitter mode, as in Fig. 5–15, the output pulse is inverted in polarity with respect to the input pulse. On the other hand, if the switching transistor is operated in the common-base or common-collector mode, the output pulse is not inverted in polarity. All three modes of operation provide gain. However, the common-collector mode provides current gain and power gain; it does not provide voltage gain. As noted in the first chapter, the input and output impedances of a transistor are quite different in the three modes of operation.

The cutoff region in Fig. 5–17(a) includes the area above the zero base-current curve ($I_B = 0$). Theoretically, with no initial base current there would be zero collector current. In turn, the collector potential would equal the battery voltage V_{cc} in Fig. 5–15(a). However, at point X on the load line in Fig. 5–17(a), a small amount of collector current is measured. This is more clearly indicated in Fig. 5–17(b) where at zero bias voltage (point X) the collector current I_c equals approximately 0.05 mA. This is the reverse-bias collector current for the *CE* configuration. Note that the application of a small reverse bias voltage V_{BE} (approximately 0.075 volt, positive for a PNP transistor), reduces the value of reverse-bias collector current to the value of I_{CBO}. The collector voltage V_{CE} is indicated by the vertical projection from point Y in Fig. 5–7(a) to the collector-voltage axis. This value is equal to the difference in magnitude between the battery voltage (12 volts in this example) and the voltage drop produced by reverse-bias collector-current flow

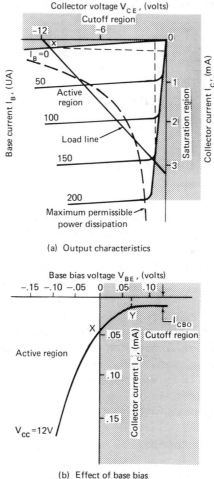

(a) Output characteristics

(b) Effect of base bias
voltage on collector current

Fig. 5–17. Transistor characteristic curves in the switching application.

through load resistor R_L [Fig. 5–15(a)]. Normal quiescent conditions for a transistor switch in this region require that both emitter-base and collector-base junctions must be reverse-biased.

With switch S1 [Fig. 5–15(a)] in its off position, the emitter-base junction is reverse-biased by battery V_{B2} through resistor R_{B2}. This is comparable to the application of a positive unit-step voltage. The collector junction is reverse-biased by battery V_{CC} through load resistor R_L; the transistor is in its *off* (cutoff) state. Next, consider the saturation region in Fig. 5–17(a). An increase in base current *does not* cause an appreciable increase in collector current I_c. At point Y on the load line, the transistor is in its saturation region. Collector current I_c is at a maximum, and collector voltage V_{CE} is at a minimum. This value of collector voltage is referred to as the *saturation voltage* (V_{SAT}), and is a basic consideration in switching function. When collector current I_c reaches its limited value (battery voltage V_{CC} divided by R_L), the transistor saturates and is in its *on* state. In this state, the transistor is equivalent to a closed switch.

5.9   TROUBLESHOOTING TELEPHONE SYSTEMS

Troubleshooting procedures for telephone systems can be basically grouped under line faults and equipment faults. There are many causes of line faults. For example, poles that support cables or open-wire lines may be struck by automobiles, resulting in mechanical damage. Lightning may strike a line during thunderstorms. When power lines are supported above telephone lines on the same poles, a break in a power line can cause burn-out of the telephone line. Vandals occasionally shoot at exposed cables or insulators. Trees often grow into both telephone and power lines and cause faults. During hurricanes, lines may be blown down or struck by falling trees. In icy weather, heavy accumulations of sleet may load the lines past their breaking point. In locations exposed to salt spray, metallic corrosion and insulator leakage can become troublesome.

When an open-circuit or a short-circuit occurs in a line or cable, it is necessary to locate the fault as accurately as possible by means of electrical measurements. Telephone technicians employ bridge-type equipment for this purpose. Figure 5–18 shows the basic arrangement of a Wheatstone resistance bridge connected to a line that has a short-circuit at a distant point. If the fault itself

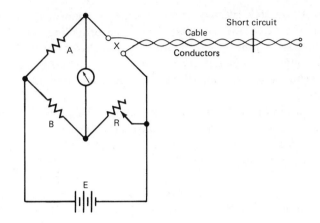

Fig. 5–18.   Wheatstone bridge connected to measure line resistance.

has zero resistance, it is easy to locate the fault. In other words, the measured resistance value is equal to the loop resistance of the pair of cable (or line) conductors from the point of measurement to the point of fault. Since the resistance per mile of the conductors is known, the distance to the fault can be determined by simple arithmetic.

If the fault shown in Fig. 5–18 has an appreciable resistance value, as is often the case, it is evident that the resistance value measured by the bridge will be equal to the loop resistance of the conductors plus the resistance of the fault. To determine whether the fault has zero resistance or some appreciable resistance value, two measurements are made, the first with the far end of the conductors short-circuited and the second with the far end open-circuited. In turn, if both measurements yield the same value of resistance, it is indicated that the fault is a *dead* short-circuit. Or if the two measurements differ in resistance values, it is indicated that the fault is a resistive-type of defect.

When the fault is resistive, another pair of measurements is required to localize the fault. A resistance measurement is made from each end of the line. In turn, a hypothetical location of the fault is made on the basis of each measurement, with the (expedient) assumption that the fault has zero resistance. These calculations lead to two different distance values and the actual location of the fault is half-way between the two hypothetical fault locations. When it is impractical to bring the bridge to the distant end of a line, it is often a simple matter to connect the distant end of the defective line to a good line, thus providing a return circuit to the testing point, as shown in Fig. 5–19. Note that if the length of the good line is not known, it can be quickly determined by a resistance measurement with the line short-circuited at the distant end.

Resistance measurements can also be used to locate a defect caused by a grounded open wire or cable conductor. The Varley loop test depicted in Fig. 5–20 is widely used for this purpose. The variable resistance R is placed in series with the resistance d of the defective conductor from the point of test to the fault. The resistance denoted as X in Fig. 5–19 becomes in Fig. 5–20 the series resistance 1 of the normal conductor plus the resistance x of the defective conductor from the far end to the fault location. The battery connection is made through the ground to the fault. This connection method avoids any error in resistance measurements caused by variation in ground resistance. When a bridge balance is obtained, the value of R is equal to the loop resistance of the circuit from the fault to the distant end (assuming that the A and B arms of the bridge are equal). In turn, it is a simple arithmetical problem to calculate the location of the fault from the far end of the line.

When the fault is a open circuit, resistance measurements cannot be employed. Instead, a pulse test may be used. The output from a pulse generator is applied to the end of the line with an

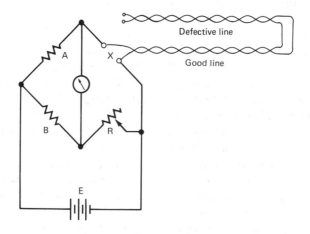

Fig. 5–19.   Return line connected to facilitate measurements.

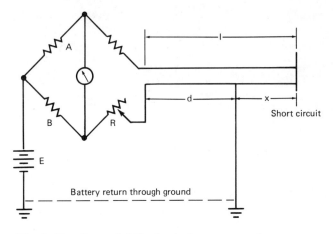

Fig. 5–20.   Grounded Varley test arrangement.

oscilloscope connected in parallel. When the pulse is applied to the line, a high-amplitude pulse is displayed on the screen of the oscilloscope. A short time later, a lower-amplitude pulse is displayed to the right of the first pulse. This is a reflected pulse, produced by electrical energy being turned back at the open-circuit point. A triggered-sweep oscilloscope is used to measure the precise time between the applied and the reflected pulses. Since the velocity of wave travel on a line or cable pair is known, it is a simple matter to calculate the distance to the open circuit from the elapsed time between the two pulses. Note in passing that the velocity of wave travel on an open line is nearly equal to the velocity of light and that a cable imposes a lower velocity depending upon the type of dielectric separating the conductors.

Equipment faults involve the same general approaches and procedures that have been discussed in earlier chapters. However, a telephone unit may be vastly more complex than an intercom or PA unit, for example. Therefore, the telephone technician requires special training and familiarization with the equipment to which he is assigned. No one person can hope to become competent in troubleshooting all types of equipment used in a large telephone system. Therefore, the system is subdivided into various areas such as mechanical switching, electronic switching, repeater troubleshooting and repair, distributing frames, power installations for central offices, and so on. The telephone troubleshooter is a specialist and he tends to become more highly specialized as the sophistication of telephone systems steadily increases.

6

THEATER
SOUND SYSTEMS

6.1   GENERAL CONSIDERATIONS

Theater sound systems differ considerably from the audio systems
previously discussed. The chief distinction is in the method of
sound recording and reproduction. For theater systems, the audio
signal is recorded on film and is reproduced by means of a light
beam and photoelectric cell. The sound may be recorded on the
same film as the picture or it may be recorded on a separate film.
In most cases, the recording consists of an optical band from 0.1 to
0.2 inch in width, running at the right-hand side of the picture and
next to the sprocket holes in the film. When wide screens were
introduced in 1952, the traditional set of speakers behind the screen
were supplemented by a speaker system placed at various locations
in the audience area so that a sound passage could always be
radiated from a realistic source. Thus, the new technique antic-
ipated quadriphonic sound. The Reeves system employed seven

sound channels. The Cinemascope system, introduced in 1953, employed a four-channel sound track utilizing a magnetic strip on the film. In 1954, MGM introduced an optical three-channel sound system with subsonic cuing signals to switch various groups of speakers on and off.

6.2   SOUND TRACK

A sound track is a narrow strip of exposed film beside the image on a motion-picture film, as illustrated in Fig. 6–1. The density of the track varies in accordance with the amplitude of the recorded audio signal. Although persistence of vision is adequate at an exposure rate of 16 frames per second, a film with a sound track is exposed at the rate of 24 frames per second. This increase in film speed permits recording of comparatively high audio frequencies. Original picture and sound *takes* are general recorded on separate strips of film to facilitate editing of both the picture and sound material. In the final stage of production, the sound track with its changes and dubs is exposed on a film strip along with the edited picture frames. It is positioned 15 inches ahead of the corresponding pic-

Fig. 6–1.   Two frames of a motion-picture film with a sound track. (Courtesy of Bell Telephone Laboratories, Inc.)

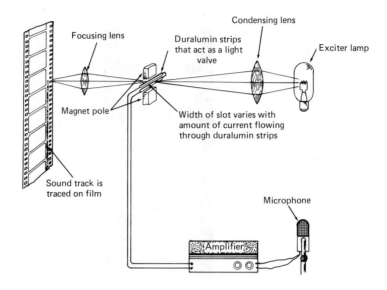

Fig. 6–2. Basic method of recording a sound track on film.

ture image. In turn, a free loop with constant velocity passes through the sound transducer while the picture images proceed in sudden jerks through the projection lenses.

Basically, sound is recorded on film as depicted in Fig. 6–2. Audio signals from the microphone are stepped up by an amplifier and are passed through a pair of metal strips that vibrate between the poles of a permanent magnet. These metal strips operate as a light valve, with a slit approximately 0.001 inch wide. When the audio-frequency currents flow through the strips, the magnetic field exerts a force on the strips and causes the width of the slit to vary accordingly. The beam of light passing through this slit varies in intensity, thereby exposing light and dark bands along the film track. These bands appear as shown in Fig. 6–3, and form a *variable-density* sound track. Next, in the reproduction of the signal by the projector, an arrangement is employed as depicted in Fig. 6–4. A beam of light is passed through a narrow slit and focused on the sound track and then into a photoelectric cell. Variations of light intensity in the photocell cause corresponding variations of current flow through the cell and reproduce the audio signal. Typical photocells used in this application are illustrated in Fig. 6–5.

Instead of employing a variable-density sound track, a *varia-*

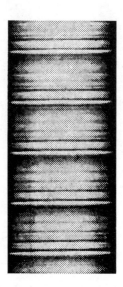

Fig. 6–3. A variable-density sound track. (Courtesy of Bell Telephone Laboratories, Inc.)

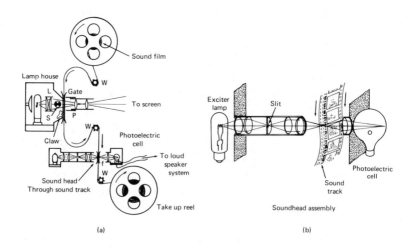

Fig. 6–4. Plan of a sound-track playback system. (a) Sound track is played back at a free loop. (b) Playback optical system.

Fig. 6–5. Typical photocells used in motion-picture projectors. (Courtesy of RCA)

ble-area sound track is often used, as exemplified in Fig. 6–6. There are two basic types of variable-area sound tracks, called the unilateral and bilateral types. A unilateral track has an unsymmetrical form with tooth-like projections of various sizes. On the other hand, a bilateral track has a symmetrical form that varies in area from one point to another. Original recordings are often made with push-pull sound tracks of the variable-density or the variable-area type, as exemplified in Fig. 6–7. This type of sound track is twice as wide as a single-ended sound track. The chief advantage of a push-pull

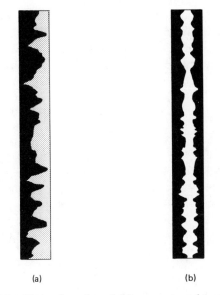

(a)　　　　　　　　　　(b)

Fig. 6–6. Examples of variable-area sound tracks. (a) Unilateral type. (b) Bilateral type. (Courtesy of Eastman Kodak Co.)

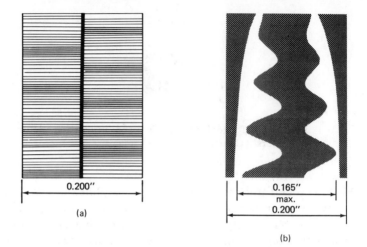

0.200"

(a)

0.165"
max.
0.200"

(b)

Fig. 6–7. Push-pull sound tracks used in original re-cordings. (a) Variable-density type. (b) Variable-area type.

track is in the cancellation of even harmonics, allowing higher fidelity to be obtained. In the final stage of production, the push-pull track is recorded as a single-ended track on the films that are released for exhibition. Push-pull sound tracks are reproduced by means of a double-cathode photocell, as shown in Fig. 6–8. The output from the push-pull transformer is a single-ended audio signal.

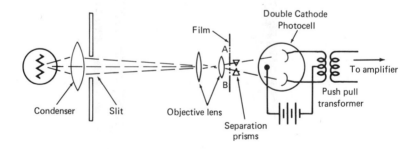

Fig. 6–8. Sound reproduction from a push-pull sound track.

6.3　PREPARING THE SOUND TRACK

Sound pickup equipment used in producing the original sound track is quite similar to the equipment employed in radio broadcasting. In the production shooting process, the microphone is operated at the end of a boom that is moved to follow the sound source while staying out of the visible scene being photographed. In case that objectionable noise occurs during exterior shooting, this portion of the sound track is edited out and a new sound track synchronized to the picture action is produced on a stage. Although the sound track is recorded on the same film with the picture for news reels, studio recording is made on a separate film. Film speed in a 35-mm recorder is approximately 90 ft/min. The picture and sound films are synchronized by means of a synchronous motor drive, or by a selsyn system. A comparatively high speed of film travel permits high-fidelity recording and provides a margin for deterioration in subsequent sound-track processing procedures.

Prints are made from the original negative sound recording, and are edited in synchronism with the edited picture. The film is then reviewed for needed cuing of music and sound effects. A special scoring stage is employed to record the music for cuing with the microphone placed for optimum pickup. This recording is carefully prepared for synchronization with the conductor viewing the picture while the music is played under his direction. A vocal number is often prescored and recorded before production shooting. It is then played back on the set with the actors *mouthing* the lines. Sound effects are recorded on separate sound tracks and synchronized with the picture. In the rerecording room, the separate sound tracks are controlled in volume and tone, edited, and a composite recording is finally made. This sound track is used for release printing.

Electronic mixers, limiting amplifiers, or limiters are generally utilized to prevent overloading on strong signal peaks, as explained previously under broadcast audio equipment. Acetate recordings are often made simultaneously with sound-track recordings so that an immediate playback can be made whenever it is desired to check on sound quality. This is particularly advantageous while recording musical passages (scoring). Figure 6–9 depicts the arrangement of a dialogue or scoring channel. Two micro-

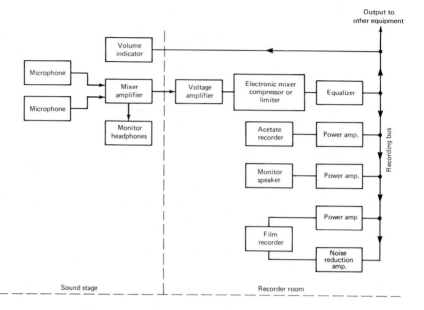

Fig. 6–9.   Arrangement of a dialogue or scoring channel.

phones are employed for dialogue recording and additional microphones are frequently used during scoring. Dialogue microphones
are placed not more than four feet away from the actors. Unidirectional microphones are commonly chosen to minimize noise pickup.
A noise-reduction amplifier is generally included in the recorder
room also, as explained in greater detail shortly. The acoustics of
the sound stage for the recording channel are carefully designed for
optimum reverberation characteristics and double-wall construction with air spaces between walls is often used to minimize the noise
level.

The microphones feed into a mixer amplifier (Fig. 6–9) that
is located at a point where the control operator will have a clear view
of the actors. The operator wears a pair of high-fidelity headphones for monitoring the audio signal. He also has a volume-
indicator (VU) meter before him that indicates the level of the
mixed output. From the mixer amplifier, the audio signal proceeds
to a voltage amplifier located in a recorder room. This room is
sometimes on or near the sound stage, but it is often placed in a
recording truck adjacent to the stage or even in a building with
cable interconnections elsewhere on the studio grounds. From the

voltage amplifier, the audio signal is fed to a limiting amplifier or equivalent unit, and then to an equalizer which maintains a uniform system frequency response, as explained previously.

From the equalizer (Fig. 6–9), the audio signal is applied to the recording bus, which is connected to an acetate recorder section, a monitor speaker located in the recorder room, the film-recorder section, and sometimes to additional equipment also. A noise-reduction amplifier is generally used in the film-recorder channel, as depicted in Fig. 6–10. This is a system that effectively lowers the level of background noise during the time that the noise is not masked by comparatively loud sounds. Its operating principle exploits the fact that the background noise from a light print is greater than that from a dark print in a variable-density sound track. In operation, the average exposure of the negative is reduced during periods of low modulation. One widely used system employs a light valve in which the spacing between the metallic strips varies in average value from 0.001 to 0.0003 inch as the audio signal decreases in average amplitude. As seen in Fig. 6–10, an elaborated gain-control channel is utilized. The timing filter is designed to increase the light-valve aperture rapidly and to decrease it at a comparatively slow rate. The other blocks operate to step up, rectify, and filter the control voltage.

As noted before, recordings from dialogue or scoring channels are subsequently rerecorded. This process is carried out with the

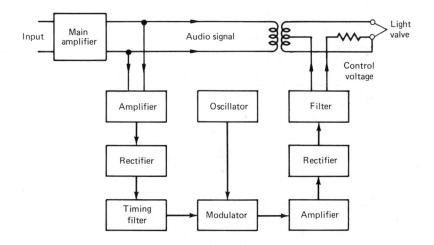

Fig. 6–10.   Block diagram of a noise-reduction channel.

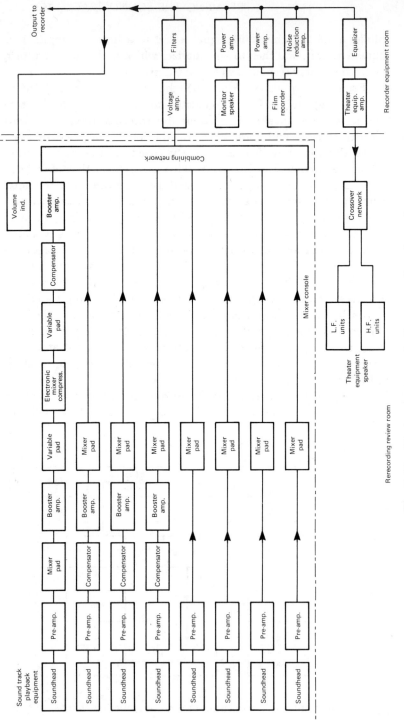

Fig. 6–11. Basic arrangement of a rerecording channel.

164

typical equipment shown in Fig. 6–11 located in a room that has acoustic characteristics similar to those of a theater. In turn, the monitor operator can determine the optimum sound quality to use in rerecording the audio signals on the master sound track. To repeat, the various audio signals are generally recorded on individual films, and include dialogue, musical, and sound-effect tracks. Each signal is separately controllable at the mixer console, and selsyn motors are used to maintain precise synchronism of the various sound tracks. Controls are provided at the mixer console for adjustment of level, frequency response, and special effects such as reverberation, telephone-line reproduction, old-time radio or phonograph reproduction, and so on. Special effects are accomplished by means of filters and various types of compensators. Reverberation is often obtained by means of a speaker in a reverberation chamber with a microphone at the other end of the chamber.

As shown in Fig. 6–7, original recordings may be of the push-pull type. This is advantageous in newsreel production, for example, because noise reduction can be realized in class *B* push-pull recording without adding the weight of a noise-reduction amplifier to the recording camera. Since push-pull sound tracks are not accommodated by theater projectors, the push-pull tracks must be rerecorded to standard single-ended track form. However, this requirement does not increase production costs because a newsreel is commonly accompanied by sound effects and various musical accompaniment. Thus, rerecording is generally specified regardless of the types of original sound tracks made on location. Most rerecordings, whether of studio scenes or news events, are released with less than high-fidelity characteristics. A frequency response from 65 to 7500 Hz is typical and limiting amplifiers are generally operated with a dynamic range of approximately 25 dB. These limits represent a trade-off between satisfactory sound reproduction and inherent production costs.

6.4   THEATER SOUND REPRODUCTION SYSTEMS

A sound track is played back in a theater by means of a soundhead, as noted previously. Figure 6–12 shows how two soundheads are employed so that little or no interruption of a program occurs when changing from one reel to another. The audio outputs from

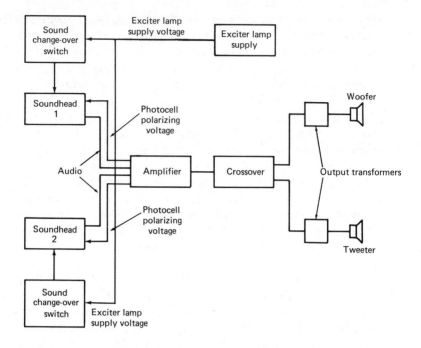

Fig. 6–12. Block diagram of a theater sound system.

the soundheads are fed to an amplifier and then to speakers that are customarily mounted behind the screen. In addition, speakers may be mounted on either side of the screen. The amplifiers are essentially the same as those used in high-fidelity systems and their power output depends upon the size of the theater. It is standard practice to utilize at least one woofer and one tweeter behind the screen. In addition a woofer and a tweeter may be installed on either side of the screen, to enhance the volume of dramatic sound passages. These auxiliary speakers are switched into or out of the circuit by means of a control track. This control track consists of variations in exposure of the small film areas between adjacent sprocket holes. An electronic switch is thereby energized to control the auxiliary speakers.

Horn-type speakers are generally used in theater sound systems. A horn speaker provides increased operating efficiency and ample space is available for installation behind the screen. As exemplified in Fig. 6–13, multicellular horn tweeters are commonly employed with a pair of logarithmic horn woofers. The woofers are much larger than those in high-fidelity systems. A uniform fre-

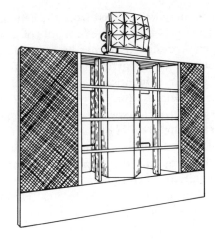

Fig. 6–13. A typical theater woofer-tweeter arrangement.

quency response is provided from approximately 70 to 7000 hz. Similar woofer-tweeter auxiliary speakers may be installed on either side of the screen. Earlier arrangements in which supplementary speakers were installed at various other locations in the theater have been generally abandoned. It is the general concensus that theater sound is most realistic when its origin is close to the screen.

6.5   THEATER ACOUSTICS

Theater acoustics are an essential component of the sound system. For example, everyone has experienced the difference between the sound of an empty house and of a furnished one. The acoustic environment of a theater is controlled by various design features. Among these, the proper use of sound-absorbing materials is important. Since people absorb sound extensively, a fully occupied theater has a shorter reverberation time than a nearly empty theater. This variable factor can be compensated for by the installation of upholstered seats; a fully-upholstered seat absorbs almost as much sound when unoccupied as when occupied. It is good practice to design the ceiling of a theater as a sound mirror in order to bring reflected sound down on top of the audience, thereby reinforcing the direct sound from the screen. In large theaters, more sound energy may be received from the ceiling than from the direct source. Figure 6–14 depicts the principle of a sound mirror.

A theater is designed primarily for good speech reproduction. Speech consists of an articulated flow, of vowel and consonant sounds. These combinations are clustered around predominant tones, called *formants*, that are characteristic of the actor. The chief difference between the speaking voices of male and female performers is the pitch region of the formants. Formants are perceived most clearly in the vowel sounds. Consonants are generally of a transient character and the dominant factor in speech intelligibility is the recognition of consonant sounds. The associated audio frequencies are in the range of 750 to 8000 Hz. Therefore, intelligibility suffers more from poor high-frequency response than from poor low-frequency response. Another important factor is reverberation time. A long reverberation time reduces intelligibility because the sound of each syllable becomes obscured by reverberation of previous syllables. An average reverberation time of 1 second may be noted; however, a shorter reverberation time is optimum for a very small theater and a longer reverberation time is optimum for a very large theater.

Reflecting areas should be designed at angles that reflect sound to the distant portion of the audience. The design technique must be directed to reflection without focusing and avoiding echoes. When plastered walls are used, a suitable reverberation time is obtained by designing the room for a volume of approximately 130 cubic feet per seat. On the other hand, when other wall substances are utilized, the optimum volume figure is changed. If the reverberation time is held constant and the volume of the room is increased, articulation decreases. In other words, interacting acous-

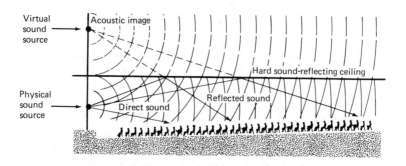

Fig. 6–14.   Basic principles of a sound mirror.

tical factors must be brought into optimum balance when designing a theater. Intelligibility improves as the loudness level increases
up to the point when the sound becomes uncomfortable and un-
natural. The amount of direct sound that reaches the distant audience increases as its elevation is raised (see Fig. 6–14). Of course,
there is a practical limit in this regard, and theater-goers in general
oppose Greek-theater seating arrangements. However, this limitation can be largely overcome by exploiting the virtual sound source
provided by a hard sound-reflecting ceiling.

7

ELECTRONIC ORGAN SOUND SYSTEMS

7.1 GENERAL CONSIDERATIONS

Electronic organs have undergone extensive development within the past forty years. This technology started with simple variable-frequency audio oscillators that played one note at a time. Instruments in this category are called melodic organs and are generally regarded as toys today. In turn, polyphonic (full-chord) organs were developed following pipe-organ characteristics. Among these are Hammond, Compton Electrone, Hoschke, Everett Orgatron, and Wurlitzer. Pioneer designs employed various types of tone generators (some of the electro-mechanical variety) and electron-tube amplifiers, as described shortly. With the advent of the transistor and of general semiconductor technology, electron tubes became obsolete; all current organs employ semiconductors. Computer technology has contributed to organ design and diode gates have replaced various mechanical switches. Today, little attempt is made to imitate pipe-organ tonal quality. In other words, the electronic organ is regarded as a distinct musical instrument. Many tonal timbres (voices) that are not within the capability of a pipe organ are provided by electronic organs. But a pipe organ can produce massive bass tones that are beyond the capability of an electronic organ.

7.2   BASIC ELECTRONIC ORGAN FUNCTIONS

A typical electronic organ is illustrated in Fig. 7–1. A manual or clavier is any keyboard or pedal board operated by the hands or feet. An expression pedal operates as a volume control. Various voices are produced by mixing a fundamental tone with various

Fig. 7–1. Appearance of a typical electronic organ. (Courtesy of Heath Co.)

proportions of harmonics. Percussion denotes characteristic tones, for instance, those produced by plucking or striking strings. Repeat percussion consists of automatic repetition of the tone as long as the tab (tablet) switch is turned on. Figure 7–2 depicts the block diagram of a typical electronic organ. In this example, separate oscillators are utilized for production of each voice, such as celeste, diapason, reed, and flute voices. Note that a swell manual (solo manual) is the upper manual of an organ; the great manual (accompaniment manual) is a lower manual used for playing the accompaniment to a melody. Some organs provide two great manuals in addition to the swell manual.

A coupler is a stop tab that permits the tones on one manual to be played with the keys of another manual or that permits the sounding of octavely-related tones on the same manual. A slide coupler is operated by pushing it in or out. Chiff denotes an accentuation of the harmonics in a voice tone during the brief period that it builds up to full amplitude after being keyed on. The purpose of chiff is to obtain a simulation of pipe-organ transient characteristics. A Chinese-block oscillator produces a novelty type of percussion tone. Note that three speakers are utilized in the example of Fig. 7–2. Two of the speakers are fixed in position, and serve as a woofer and a tweeter. On the other hand, the gyrophonic speaker consists of two (or sometimes three) speakers mounted on a rotating disc. Each of the moving speakers reproduces a separate voice of the organ. Rotation of the gyrophonic speakers with respect to the fixed speakers produces a periodic phase variation. This is heard as an ensemble or choral effect. Unless there is phase variation among the voices in an ensemble, a choral timbre is lacking.

A gyrophonic speaker is a method of obtaining mechanical vibrato. The phase variation occurs at a rate of approximately six times per second. Mechanical vibrato is also called *theatrical tremolo*. A phase variation is equivalent to a frequency variation. Vibrato is defined as frequency modulation, whereas tremolo is defined as amplitude modulation. Many organs provide a 7-Hz amplitude modulator to add a tremolo timbre to various voices. When a key is depressed on a manual, the selected voice is said to speak. Another type of speaker that provides mechanical vibrato is the Badwin-Leslie type, depicted in Fig. 7–3. It employs a diagonally-mounted rotating disc beneath the speaker that deflects the sound around a circular arc seven times per second. In turn, a cyclic Doppler effect results, which is equivalent to a frequency modula-

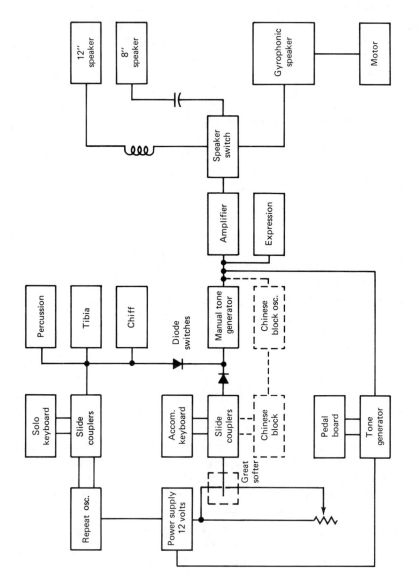

Fig. 7–2. Block diagram of a typical electronic organ.

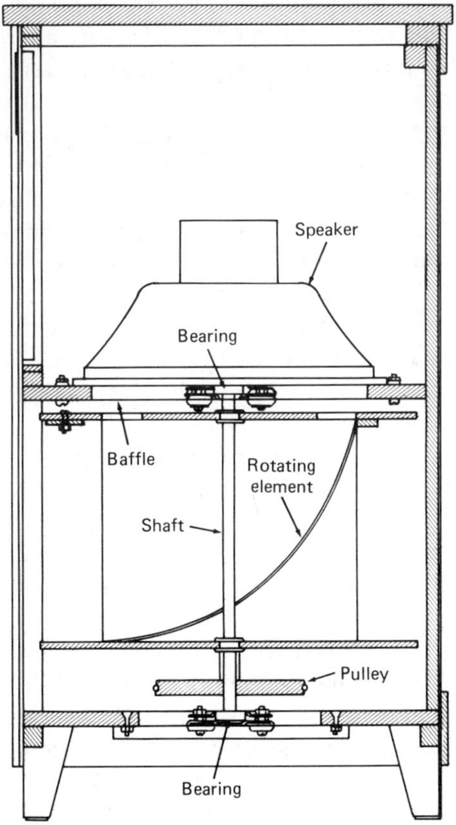

Fig. 7–3. Plan of the Baldwin-Leslie theatrical-vibrato speaker. (Courtesy of Baldwin Organ Co.)

tion of the tone. The timbre that is provided resembles that from a gyrophonic speaker. Organ speakers are mounted in enclosures which are called tone cabinets. Note that an organ audio system is basically different from a high-fidelity system, in that an organ is designed to *produce* musical tones whereas a high-fidelity system is designed to *reproduce* musical tones.

7.3 ORGAN TONE GENERATORS

Tone generators are basically audio oscillators. In the past, sine-wave oscillators have been employed. However, it is now standard practice to utilize complex-waveform oscillators. No musical tones are pure sine waves, although a flute note approaches a sine waveform. When sine-wave tone generators were used in electronic organs, musical tones were synthesized (built up) by the addition of numerous harmonics to a fundamental frequency. This method is theoretically correct but its disadvantage is that prohibitively complex circuitry must be used to provide other than the most basic voices. Therefore, modern organs produce musical tones by shaping complex waveforms as required with low-pass, high-pass, and bandpass filters. These formant filters are explained in greater detail later. Many tone generators are essentially sawtooth waveform generators. A sawtooth waveform has an advantage over a square waveform, for example, in that the sawtooth has both even and odd harmonics.

A basic sawtooth generator configuration is shown in Fig. 7–4. This is a blocking-oscillator arrangement with a shunt capacitor C2 in the output circuit. A semi-sawtooth waveform is provided that can be filtered in various ways to produce different musical tones. The harmonic composition of a sawtooth waveform is depicted in Fig. 7–5. Successive harmonics have amplitudes which are inversely proportional to the order of the harmonic. Note that if individual sine-wave oscillators are used to synthesize a sawtooth waveform, many oscillators are required. On the other hand, a sawtooth waveform can be generated with a single blocking oscillator. Alternatively, a sawtooth waveform can be obtained by differentiating the output from a rectangular-wave oscillator. Note in passing that some tone generators have output waveforms that cannot be classified in one of the basic categories. Any complex wave-

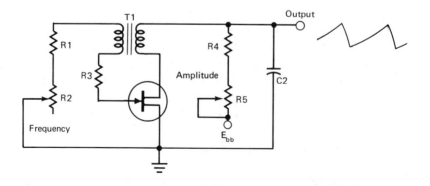

Fig. 7–4. Basic sawtooth generator configuration.

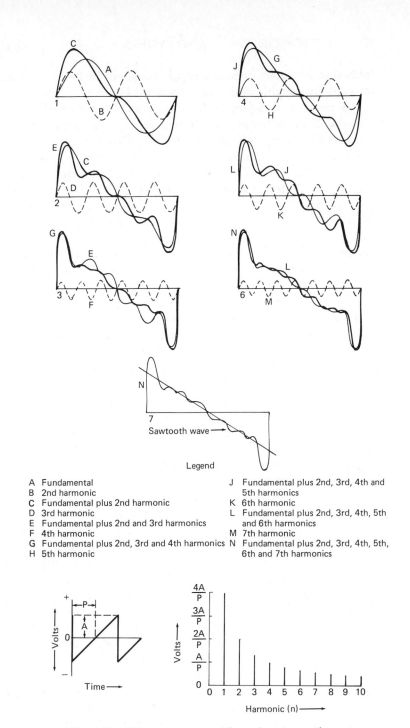

Legend

A Fundamental
B 2nd harmonic
C Fundamental plus 2nd harmonic
D 3rd harmonic
E Fundamental plus 2nd and 3rd harmonics
F 4th harmonic
G Fundamental plus 2nd, 3rd and 4th harmonics
H 5th harmonic

J Fundamental plus 2nd, 3rd, 4th and 5th harmonics
K 6th harmonic
L Fundamental plus 2nd, 3rd, 4th, 5th and 6th harmonics
M 7th harmonic
N Fundamental plus 2nd, 3rd, 4th, 5th, 6th and 7th harmonics

Fig. 7–5. Harmonic composition of a sawtooth waveform.

177

form that has both even and odd harmonics can be utilized satisfactorily by suitable design of subsequent circuitry.

Some organs employ an individual oscillator for each note. On the other hand, many organs utilize a single set of master oscillators for the top octave, with frequency dividers to develop the frequencies for the lower octaves. It is a considerable design advantage to use 12 oscillators followed by frequency dividers instead of 120 individual oscillators, for example. However, there is a basic disadvantage in the frequency-divider system in that corresponding notes of different octaves are phase-locked. In turn, chords tend to lack full musical richness. To minimize this disadvantage, various expedients are employed to enrich the choral and orchestral effects of phase-locked notes. Previous mention has been made of the fixed and moving speakers supplied by different organ voices.

A typical frequency-divider network is depicted in Fig. 7–6. The two divider sections in this example employ packaged electronic circuits (PECs) and the master oscillator employs conventional printed-circuit construction. Three output frequencies are provided by this network that are integrally related in 4–2–1 values. The divider output waveforms are essentially square waves. Twelve master oscillators and 48 divider sections are used. The divider output waveforms are mixed as exemplified in Fig. 7–7 to produce semi-sawtooth (staircase) waveforms. Note that a square wave has odd harmonics only, whereas a staircase waveform has both even and odd harmonics. It is necessary to generate both even and odd harmonics since both are required to form various musical tones.

It is a basic law of waveshaping that no harmonics can be produced by a passive network. Only the relative amplitudes of the harmonics can be changed by a differentiating circuit, integrating circuit, or filter configuration. To introduce new harmonics into a waveform, a nonlinear device of some type must be utilized. For example, a transistor switching circuit is a nonlinear arrangement. The frequency dividers shown in Fig. 7–6 are multivibrator configurations where the transistors operate as switches. In turn, harmonics that were not present in the input waveform are introduced into the output waveform. Therefore, although there are no even harmonics in the square-wave output from the master oscillator, the first divider provides an output waveform that has a second-harmonic relation to the input waveform while the second divider provides an output waveform that has a fourth-harmonic relation to the input waveform. Consequently, the staircase waveform developed by these outputs (Fig. 7–7) contains both even and odd harmonics.

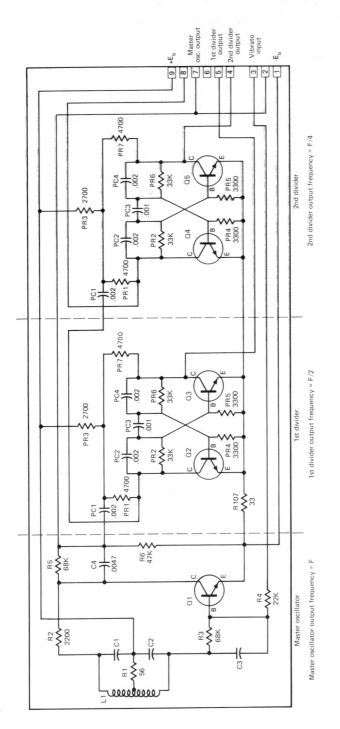

Fig. 7-6. A tone-generator and frequency-divider arrangement. (Courtesy of Heath Co.)

179

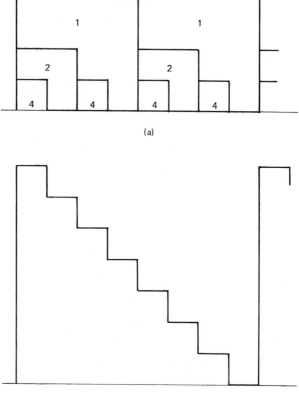

(a)

(b)

Fig. 7–7. Divider outputs are mixed to form a semi-sawtooth waveform. (a) Master-oscillator and divider waveforms. (b) Combined waveforms.

7.4 FORMANT FILTERS

Formant (voicing) filters are used to change a source waveform, such as a sawtooth, into a desired tonal waveform. Figure 7–8 shows the comparative harmonic contents of string, flute, and reed tones. Note that string and reed voices have harmonics that exceed the amplitude of the fundamental frequency while a flute tone has a dominant fundamental frequency. These various harmonic distributions are obtained by passing the basic complex waveform (such as a staircase wave) through various reactive circuits. In

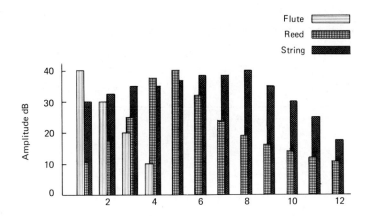

Fig. 7–8.   Comparative harmonic contents of three basic voices.

addition to simple filters, traps are sometimes used to attenuate or remove a particular group of harmonics. LCR ringing circuits are utilized occasionally to enhance a particular harmonic. If the circuit has a high Q value and rings strongly, it generates a tone that simulates a drum. This is called a *percussion waveform*. If it is automatically repeated at a predetermined rate, it is termed a *repeat-percussion waveform*.

Typical flute-voice formant filters are depicted in **Fig. 7–9.** Each section consists of a low-pass filter and a high-pass filter. This filtering action forms a distorted sine wave consisting of a fundamental with a few harmonics that are substantially attenuated. A formant filter is used for each half octave over most of the solo manual and for each octave over the accompaniment manual. Variable inductors are provided for precise adjustment of the filter frequency response. Figure 7–10 shows formant filters for solo cornet, solo trombone, vox humana, viole, geigen diapason, melodia, and accent voices. The first three are combination RC and LCR filters and the remaining four are different forms of RC filters. Tab switches for selection of various voices are shown below the filters. Note the use of a light-dependent resistor (**LDR**) for switching the spectratone signal between terminal 2 and the emitter of $Q1$. This is a form of semiconductor switching utilized in electronic organs.

A diapason voice simulates a traditional pipe-organ tone and the pitch terminology follows tradition. In other words, 8-foot voices

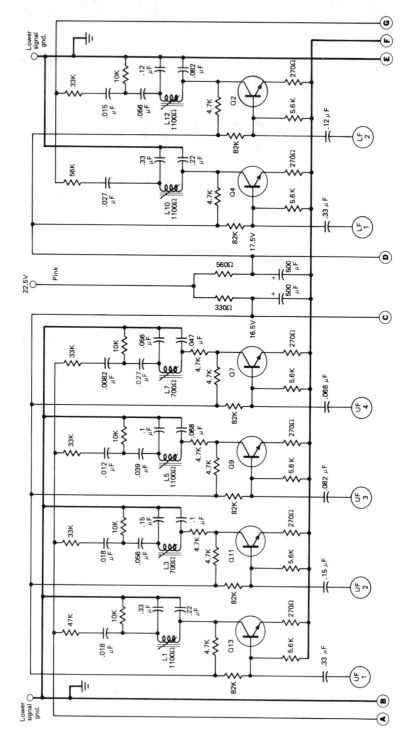

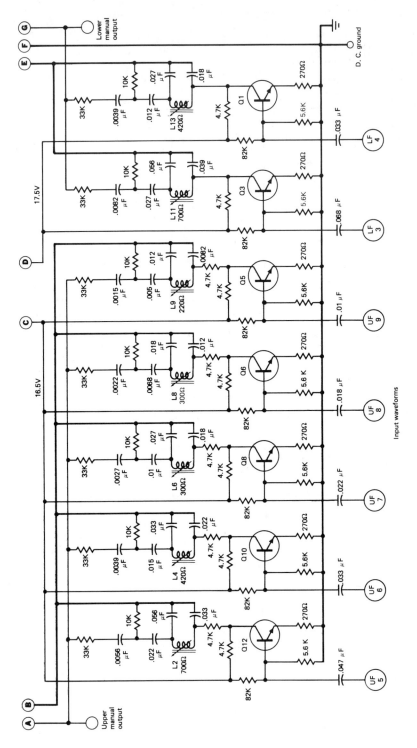

Fig. 7–9. Typical flute formant filters.

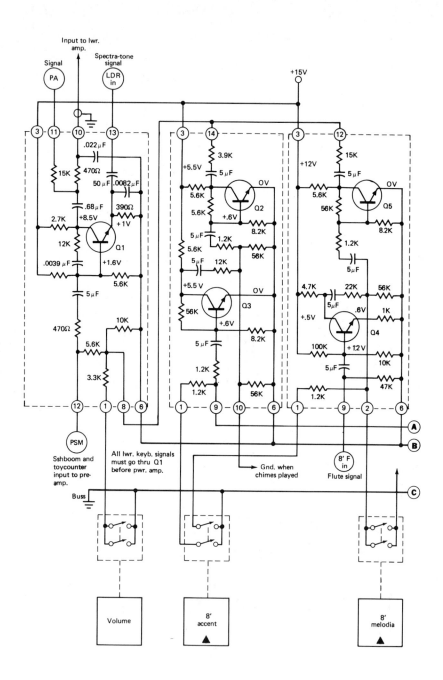

Fig. 7–10. Typical formant filters for seven basic organ voices.

184

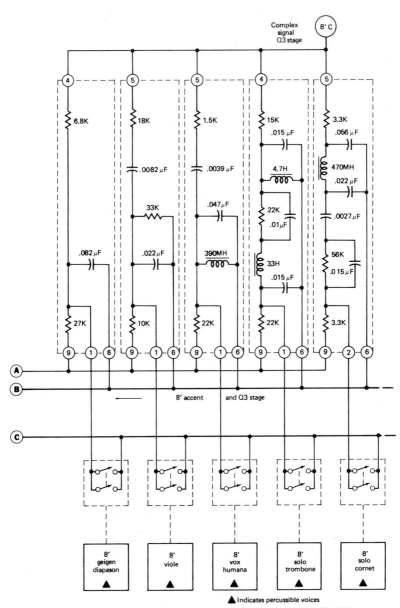

Complex
signal
Q3 stage

8' C

6.8K   18K   1.5K   15K   3.3K

.015 μF   .056 μF

4.7H   470MH
.022 μF

.0082 μF   .0039 μF

33K   .047μF   22K   .0027 μF
.01μF

.082 μF   .022 μF   390MH   56K
33H   .015 μF

27K   10K   22K   22K   3.3K

.015 μF

A

B

8' accent   and Q3 stage

C

8' geigen diapason	8' viole	8' vox humana	8' solo trombone	8' solo cornet
▲	▲	▲	▲	▲

▲ Indicates percussible voices

All switches shown in normally "off" position

are denoted in Fig. 7–10. A 16-foot audio signal has a fundamental frequency of 32.5 Hz; an 8-foot signal has a fundamental frequency of 65 Hz. Note that the 8-foot accent tab in Fig. 7–10 is not an independent voice; this tab is used merely to make a selected 8-foot voice sound louder with respect to the other footages in the total audio output. The spectratone signal is fed to a tone cabinet that produces a vibrato effect, as explained previously. To obtain a vibrato effect in the lower audio frequencies, an electronic tremulant (amplitude modulation) is provided. When used in combination with a gyrophonic speaker, choral or ensemble effects are thereby produced.

Special effects called *sshboom* and *toycounter* effects are indicated in Fig. 7–10. The toycounter function produces snare-roll and castanet-roll effects. Switching action is provided for single-percussion or repeat-percussion outputs. The sshboom effect is similar to the toycounter effect except that it produces sounds similar to cymbal and bass-drum characteristics. Typical drum frequencies are: bass drum, 100 Hz; clave, 2100 Hz; and woodblock, 900 Hz. A woodblock effect employs a shorter time constant than a drum effect. Cymbal and brush effects are produced by a random-noise generator, with filters and gates to develop the desired tone colors. A snare-drum effect combines the outputs of a percussion oscillator and a noise generator with simultaneous gating of the outputs.

7.5   MODULATION OF ORGAN VOICES

Organ voices are modulated in both steady-state and transient modes. As noted previously, a vibrato is a frequency modulation of a tone at a rate of approximately 7 Hz. A tremolo is an amplitude modulation of a tone at the same frequency. A Doppler tone cabinet (such as a Leslie speaker) produces a combination of vibrato and tremolo modulation. These are steady-state modulation processes. A wow-wow is a very slow vibrato typically produced by a configuration like that shown in Fig. 7–11. The emitter follower $Q3$ operates as a very low frequency oscillator. When the wow-wow effect is *not* being used, the wow-wow switch is in the opposite position from that shown in the diagram. Accordingly, the audio signal from the main preamplifier feeds through the expression pedal control directly to the output circuit. But when the wow-wow effect is being used, the audio signals from $Q1$ and $Q2$

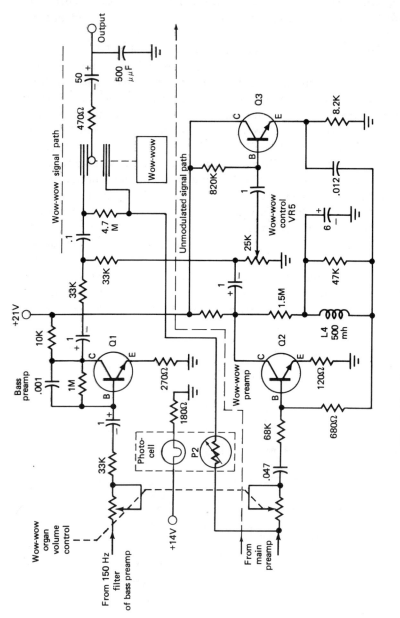

Fig. 7–11. A wow-wow modulator configuration.

are fed into the output circuit. Also, $Q3$ oscillates at a very low frequency because of feedback through the expression control. In turn, the signal passing through $Q2$ becomes modulated accordingly.

Electronic vibrato is obtained by means of a 7-Hz sine-wave oscillator, as exemplified in Fig. 7–12. The output from the oscillator modulates an amplifier following the tone generators. In turn, amplitude modulation is obtained that is a tremolo in the strict sense of the term. However, there is often little distinction made between vibrato and tremolo effects. The arrangement in Fig. 7–12 employs a phase-shift oscillator with a Darlington pair of transistors. Potentiometer $R2$ serves to adjust the percentage modulation of the audio signal. In other designs, diode modulation is employed to obtain a vibrato effect. For example, an audio amplifier transistor may utilize a forward-biased diode as an emitter resistor. Pulses with a 7-Hz repetition rate are fed into the diode thereby changing its resistance accordingly and varying the gain of the stage.

Reverberation denotes an echo effect. An organ voice acquires reverberation when it passes through a configuration such as that shown in Fig. 7–13. This arrangement comprises amplifiers and an

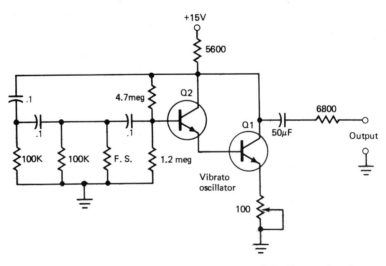

F. S. : Factory selected

Fig. 7–12. A typical vibrato oscillator configuration.

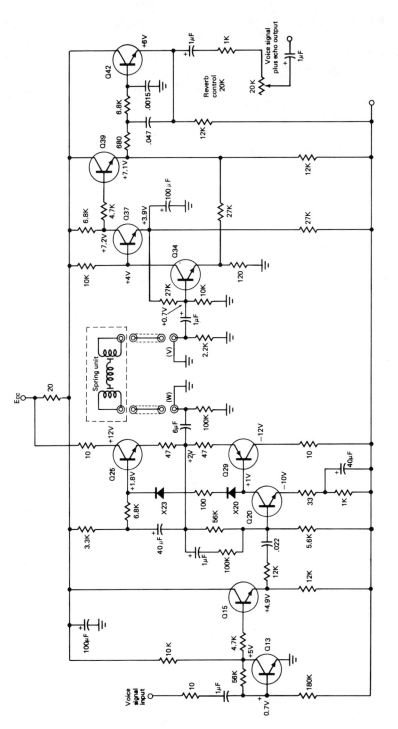

Fig. 7-13. A reverberation configuration.

electromechanical coil-spring device. The output from the first amplifier section energizes an electromagnet that produces an audio sound wave in the coil spring. At the far end of the spring, an iron armature is vibrated accordingly and induces a voltage in a pickup coil. This audio voltage is amplified and fed to the tone cabinet(s). Note that as the iron armature is vibrated, a portion of the spring energy is reflected and returns to the input end. Again, some of this energy is rereflected and arrives at the armature after a delay interval. The echo effect is obtained in this way. In most cases, several echoes are audible following a loud note.

In order to simulate pipe-organ timbres by electronic means, various expedients are employed. One of these is called *synthetic bass*; it provides an illusion of very low bass tones that are not actually present in the organ voice. To produce synthetic bass, the first few harmonics of the absent frequency are given abnormal amplitude by filter action. These harmonics beat together, and due to the nonlinear characteristic of the ear, the listener believes that the absent frequency is present. Another expedient in this category is called *chiff*. This is a transient effect at the beginning of a note that simulates wind noise from a pipe organ. Chiff is produced by introducing enhanced harmonics during the attack time of a note. As an illustration, the third and fifth harmonics may be greatly increased, while the fundamental is reduced and the second harmonic suppressed during the chiff interval. This is accomplished by means of nonlinear filter charging circuits. Figure 7–14 depicts an example of a chiff circuit.

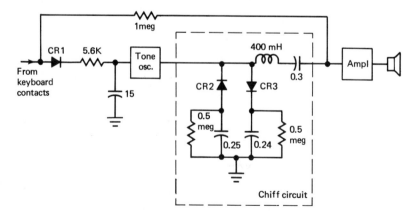

Fig. 7–14. A typical chiff generating circuit.

8

CARRIER-CURRENT AUDIO SYSTEMS

8.1 GENERAL CONSIDERATIONS

Carrier-current communication denotes information transfer by means of modulated carrier signals that are conducted from one point to another by wire lines, cables, or waveguides. Carrier communication is employed to obtain additional telephone channels on existing audio-frequency lines and circuits. It is also utilized to provide telephone channels on existing power lines that are not adapted to information transfer by means of audio-frequency currents. Note that a telephone line can be used for simultaneous transmission of audio-frequency signals and carrier currents. Thus, carrier transmission can be regarded as an evolution of the phantom circuit in developing additional communication channels. In the early stages of the art, carrier communication was termed *wired wireless*. The carrier frequency may have any value up to the cutoff frequency of the transmission line and any value down to the

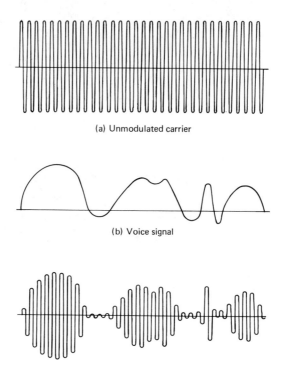

(a) Unmodulated carrier

(b) Voice signal

(c) Carrier after modulation by B.

Fig. 8–1. Modulation of a carrier by an audio signal.

limit of the audible range (approximately 12 kHz). An audio signal is amplitude-modulated on the carrier, as depicted in Fig. 8–1. Technically, the audio signal can be regarded as a modulated waveform in which the carrier frequency is 0 Hz.

8.2 CARRIER-CURRENT SYSTEMS

A 2500-Hz band of voice frequencies (200 to 2700 Hz) can be shifted by the modulation process to a 2500-Hz band of 16,000 to 18,500 Hz. Another 2500-Hz band of voice frequencies could be shifted to a band of 13,500 to 16,000 Hz and transmitted over the same line. Still another 2500-Hz band of voice frequencies could be shifted to a band of 18,500 to 21,000 Hz and transmitted over the same line. The basic plan of such a system is shown in Fig. 8–2.

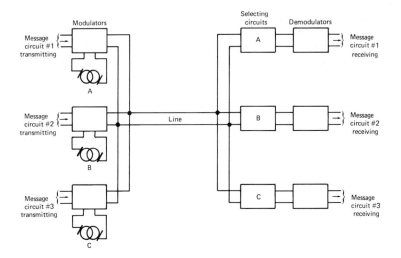

Fig. 8–2.   A basic carrier-current system.

The modulators provide the necessary band shifts and the selecting circuits (filters) sort out the individual 2500-Hz bands. In turn, the demodulators process the modulated waveforms and develop the audio-frequency envelopes for the receivers. Note that these three transmissions may occur simultaneously on the line without mutual interference.

The arrangement depicted in Fig. 8–3 provides for one-way communication only. In practice, it is generally necessary or desirable to employ two-way communication. When this elaboration is provided in the example under discussion, six communication channels are effectively available. Figure 8–3 illustrates the manner in which hybrid coils are used to obtain two-way carrier-current communication. Operation of hybrid coils was explained previously under the topic of telephone systems. The arrangement shown in Fig. 8–3(a) is suitable for general use. On the other hand, the configuration of Fig. 8–3(b) is somewhat critical with respect to balance and crosstalk; therefore, it is seldom used today.

8.3   MODULATION AND DEMODULATION

An amplitude-modulated waveform was depicted in Fig. 8–1. A basic amplitude-modulation arrangement is shown in Fig. 8–4. This

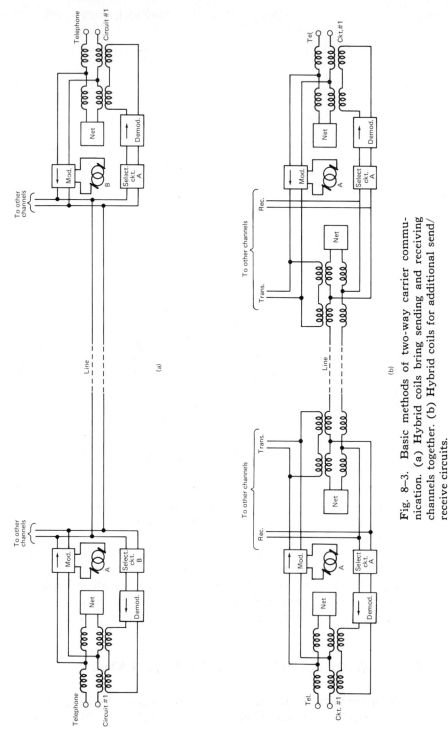

Fig. 8-3. Basic methods of two-way carrier communication. (a) Hybrid coils bring sending and receiving channels together. (b) Hybrid coils for additional send/receive circuits.

194

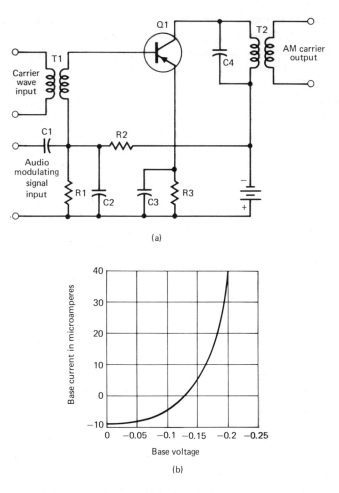

Fig. 8–4. Basic amplitude-modulation arrangement. (a) Circuit. (b) Transistor base characteristic.

is termed a *base-modulated transistor configuration* because the modulation process occurs in the base-emitter junction. Note that both the carrier-wave input and the audio-modulating signal input are applied to the base of $Q1$. In turn, the amplitude of the carrier-frequency output from the collector depends on the instantaneous value of the audio signal. This relation occurs because the base-emitter bias voltage is set to a point that provides nonlinear operation of $Q1$. For example, with reference to Fig. 8–4(b), the base-emitter bias might be set to −0.1 volt. In turn, when this bias voltage swings up and down in accordance with the audio signal, the amplification of $Q1$ is greater for an upward swing than for a downward swing. Therefore, the output voltage drop across $T2$ is an amplitude-modulated waveform.

Capacitor $C1$ in Fig. 8–4 serves as a coupling capacitor; capacitors $C2$ and $C3$ function as bypass capacitors. Capacitor $C4$ is a tuning capacitor for $T2$. Note that $C2$ must be tuned to the frequency of the carrier wave to obtain maximum signal output. Although $C2$ operates as a carrier-frequency bypass capacitor, its value is sufficiently small so that it does not bypass the audio modulating frequency. Transformer $T1$ serves as an input coupling device. Resistors $R1$, $R2$ and $R3$ operate in the base-emitter bias network. It is advantageous to bypass $R1$ in order to avoid IR signal loss; $R3$ is bypassed to avoid emitter degeneration at the signal frequency, which would reduce the stage gain. However, $R3$ provides DC emitter feedback, and thereby stabilizes the bias system.

A basic amplitude demodulator arrangement is depicted in Fig. 8–5. Diode $CR1$ is employed as a rectifier, followed by an RC filter with a short time-constant. Note that if the carrier signal input were applied to a telephone receiver, no sound would be reproduced because the average value of the waveform is zero with respect to the audio-frequency information. Therefore, the carrier signal input must be passed through a rectifier in order to eliminate one polarity of excursion. In turn, the half-wave output from $CR1$ has an average value that varies in accordance with the audio information. A telephone receiver will reproduce the audio signal when energized by this rectified waveform. However, it is advantageous to pass this rectified waveform through filter $R1$-$C1$ in order to remove the high-frequency carrier component and recover the original audio signal in pure form. In other words, the carrier component serves no useful purpose in the output circuit and may tend to overload subsequent devices.

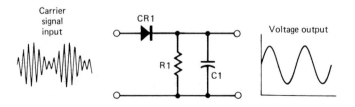

Fig. 8–5.   A basic demodulator arrangement.

Next, it is instructive to observe the frequency components of an amplitude-modulated waveform. When a carrier is amplitude-modulated by an audio frequency, a pair of sidebands is generated, as exemplified in Fig. 8–6. Thus, if a 455-kHz carrier is modulated by a 10-kHz audio signal, the output waveform consists of the original 455-kHz carrier, a 445-kHz lower sideband, and a 465-kHz upper sideband. The transmitted information is in the sidebands. Moreover, the essential information is contained in one sideband alone. This fact is illustrated in Fig. 8–7. Note that if the carrier is removed from an amplitude-modulated waveform, a double-frequency modulation envelope is produced. It follows that if the carrier is reinserted into the waveform of Fig. 8–7(b), the waveform of Fig. 8–7(a) will be recovered.

Next, refer to Fig. 8–8. When one sideband is removed from an amplitude-modulated waveform, the original modulation waveform is retained in slightly distorted form. If desired, the carrier

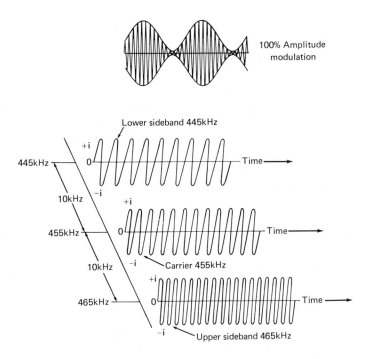

Fig. 8–6.   Frequency components of a typical modulated waveform. (a) Fully amplitude-modulated waveform. (b) Frequency components.

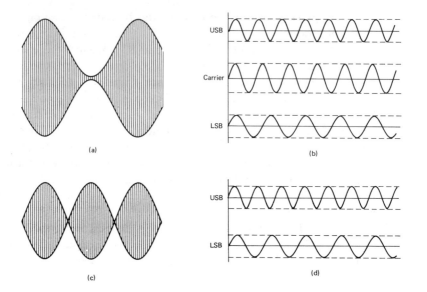

(a)

USB

Carrier

LSB

(b)

(c)

USB

LSB

(d)

Fig. 8–7. Example of carrier suppression. (a) Carrier and both sidebands present in waveform. (b) Components of waveform in (a). (c) Both sidebands without carrier in waveform. (d) Components of waveform in (c).

can be removed from a single-sideband waveform and reinserted at the receiving end of the system. There are technical advantages both in carrier suppression and in single-sideband transmission. Carrier suppression permits the information to be transmitted at a higher level and improves the signal-to-noise ratio by a factor of 9 dB. Single-sideband transmission permits information transfer in one-half the channel width that is required for transmission of a conventional amplitude-modulated waveform. For example, if a voice signal occupies a band of 2500 Hz, carrier transmission employing a single sideband will require a channel width of 2500 Hz. On the other hand, carrier transmission of the same voice signal using double-sidebands will require a channel of 5000 Hz.

Many carrier systems employ single-sideband transmission. It is instructive to note that a single-sideband waveform produced by a *steady* audio tone such as a 1-kHz frequency has a continuous sine-wave (CW) form, as depicted in Fig. 8–9. Therefore, this single-sideband signal *cannot* provide an audible output from a demodulator. On the other hand, if the missing carrier is reinserted with the SSB signal prior to demodulation, the 1-kHz beat signal

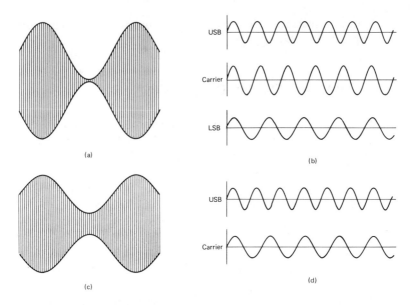

Fig. 8–8. Example of suppression of one sideband. (a) Carrier and both sidebands present in waveform. (b) Components of waveform in (a). (c) Carrier and one sideband present in waveform. (d) Components of waveform in (c).

that results becomes available in the demodulator output. Students who have had some experience with SSB reception know that a garbled and unintelligible voice output is obtained from a demodulator if the carrier is not reinserted prior to demodulation. This output results from the fact that a *voice* waveform has numerous *harmonics* (is not a steady sine-wave signal). Therefore, these harmonics produce beats that appear in the demodulator output, although they form an unintelligible sound pattern. When the car-

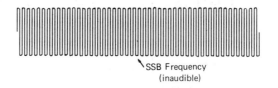

SSB Frequency
(inaudible)

Fig. 8–9. A single-sideband signal produced by a steady audio tone.

rier is reinserted, the demodulator output then represents the original modulating signal.

8.4   BALANCED MODULATION

To obtain a suppressed-carrier waveform (both sidebands minus the carrier frequency) as depicted in Fig. 8–7(b), a balanced modulator is employed, as shown in Fig. 8–10. This arrangement utilizes a pair of transistors connected to a push-pull arrangement with bridge terminals provided. Note that the carrier frequency is applied to the gates of the transistors in the same polarity. On the other hand, the audio signal is applied to the gates in push-pull. Accordingly, the transistors operate in parallel with respect to the carrier and operate in push-pull with respect to the audio signal. The output circuit is a push-pull configuration and the complete circuit constitutes a balanced bridge with respect to the carrier. Therefore, the carrier frequency cannot appear in the output. Since the transistors are biased for modulator action, amplitude modulation takes place between the gate and source of each transistor. Thus, push-pull modulator action occurs, with the carrier frequency cancelled in the output circuit. The suppressed-carrier output waveform contains an upper sideband and a lower sideband. Either of these sidebands can be removed by means of a bandpass filter with sharp cutoff characteristics.

A simple four-channel telephone system is depicted in Fig. 8–11. This arrangement uses three carrier channels in addition to a

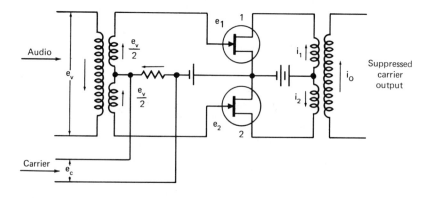

Fig. 8–10.   Basic balanced modulator configuration.

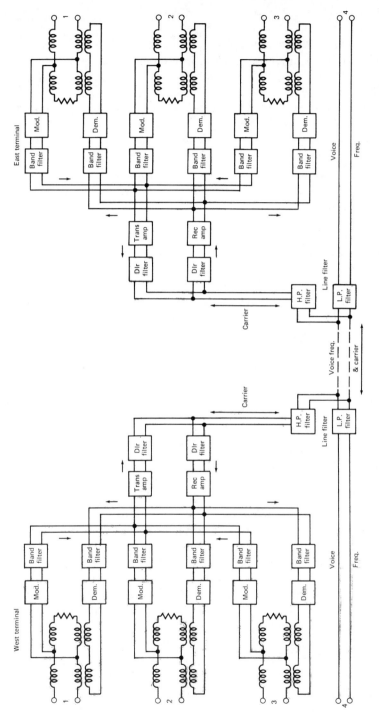

Fig. 8–11. A carrier system that provides three channels in addition to the conventional voice-frequency channel.

conventional voice channel. Each carrier channel operates in a different frequency band so that interference does not occur. Suppose that an audio signal is applied at the west terminals of channel 4. This audio signal proceeds through the low-pass filter, along the line, and through the low-pass filter at the east terminal. The audio signal is blocked from the carrier equipment by the high-pass filters. Thus, channel 4 operates as a conventional voice-frequency channel. Next, suppose that an audio signal is applied at the west terminals of channel 1. The hybrid coil feeds the signal into the modulator but prevents it from entering the demodulator. This channel employs a balanced modulator and a band filter that supplies a single sideband to the transmission amplifier. The amplified carrier signal then goes through a high-pass filter and into the voice-frequency and carrier line. Note that the carrier signal cannot flow through the low-pass filter into voice-frequency channel 4.

At the east terminal, the carrier signal flows through the high-pass filter and through the directional filter to the receiving amplifier. The signal cannot flow into the transmission amplifier through its associated directional filter. From the receiving amplifier, the carrier signal finds passage through the demodulator band filter for channel 1. In turn, the missing carrier is reinserted and the signal is demodulated in this section. From the demodulator, the audio signal flows through the hybrid coil to the output terminals for channel 1. Note that the hybrid coil prevents the audio signal from entering the modulator. Suppose that an audio signal is applied from the line to the channel-1 terminals. Now, the hybrid coil passes the incoming signal to the modulator and prevents it from entering the demodulator. The foregoing sequence of circuit actions is then repeated in the opposite direction and results in an audio-signal output at the west channel-1 terminals.

It will be observed that when an audio signal is applied to the west terminals of channel 2, a single-sideband carrier signal is generated that flows to the channel-2 east demodulator where the signal is reconstituted. In turn, the audio signal appears at the east channel-2 terminals. Since the carrier frequency is different for each channel, there is no audio output at the channel-1 or channel-3 terminals. As explained previously for channel-1 operation, there is also no output at the channel-4 terminals. In practice, the carrier frequencies are spaced 3 kHz apart. This spacing permits the use of a 2500-Hz bandwidth on each channel with a small *guard band* to ensure against interference between adjacent channels.

8.5   POWER-LINE CARRIER COMMUNICATION

Power-line carrier communication denotes telephone operation over high-voltage or low-voltage power lines. For example, a public-utility installation may involve 100,000-volt power lines while a wireless intercom installation in a residence involves 120-volt power lines. Carrier frequencies are in the range from 50 to 150 kHz. Several communication channels may be utilized simultaneously on various carrier frequencies. The arrangement of a typical power-line carrier system is depicted in Fig. 8–12. High-voltage coupling capacitors are required to prevent destruction of the carrier equipment by the transmission-line voltage. These capacitors must withstand the line voltage under all weather conditions. Capacitance values range from 0.004 to 0.00075 µf, with the smaller values employed for the higher-voltage lines. The capacitance values that are utilized represent a compromise between system efficiency, design problems, and production costs.

Refer to Fig. 8–12. The coupling capacitors are connected to a line-tuning unit. This line-tuning unit (Fig. 8–13) consists of an adjustable inductor that resonates with the capacitance of the coupling capacitor at the carrier frequency. In this manner, coupling losses are minimized and system efficiency is improved. Note that when a single-frequency system is used, only one station can talk at a time. However, the advantage to this method is that each station can listen and talk to any other station, as with a party line. The chief disadvantage of the single-frequency system is that the network connections are such that one party cannot interrupt another while he is talking. In other words, the carrier is turned on by the person who starts a call and other stations cannot turn on the carrier for their own units until the caller turns his carrier off. However, there is an automatic system that can be utilized that breaks the carrier between word pauses so that others on the *party line* can interrupt a caller.

When a two-frequency system is employed, a caller need not turn his carrier off to listen for another station and a listener can interrupt a transmission without waiting for a pause. Note, however, that the two-frequency system is limited in that only two stations can converse at the same time. Also, if half the stations in a system are tuned to operate on carrier *A*, and the other half on carrier *B*, those tuned to carrier *A* cannot communicate with one

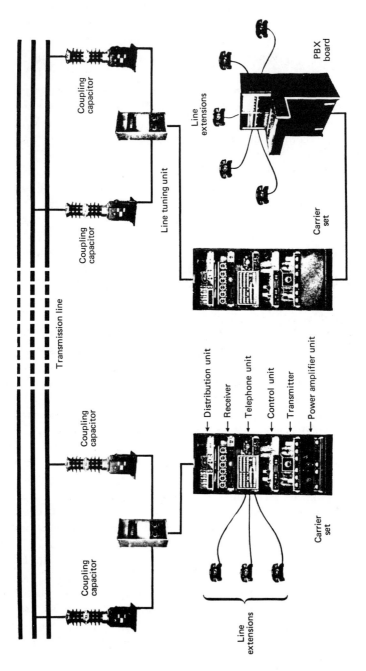

Transmission line

Coupling capacitor

Coupling capacitor

Coupling capacitor

Coupling capacitor

Line tuning unit

Line extensions

PBX board

Carrier set

Distribution unit

Receiver

Telephone unit

Control unit

Transmitter

Power amplifier unit

Carrier set

Line extensions

Fig. 8–12. Arrangement of a typical power-line carrier system. (Courtesy of Bell Telephone Laboratories, Inc.)

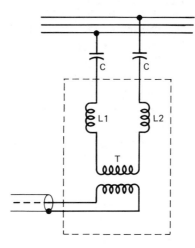

Fig. 8–13.   A basic line-tuning unit.

another at the same time. Neither can those tuned to carrier *B* communicate with one another at the same time. However, any carrier-*A* station can communicate with any carrier-*B* station at the same time. Therefore, a *party-line* installation is sometimes preferred.

The distribution unit indicated in Fig. 8–12 comprises input and output transformers. As in standard telephone systems, hybrid coils are used to prevent system oscillation (singing). A super-heterodyne carrier receiver configuration is employed with one IF stage and two audio stages of amplification. The telephone unit is essentially a dial switching arrangement. Selective ringing is provided for several handsets or extensions at both terminals. If desired, any one of the extensions can be connected into a private-branch exchange (PBX) board. The control unit comprises control relays and calling circuits. If desired, voice calling or code-bell calling can be used instead of dial selection. Loudspeakers are installed at all stations for voice calling. Code-bell calling is basically the same as automatic ringing, except that the ringing sequence is coded into long and short groups. When dial-selection calling is utilized, only a particular handset will have its bell energized.

The transmitter unit in Fig. 8–12 may be a conventional amplitude-modulated arrangement for operation over short distances. However, communication over long distances generally employs single-sideband amplitude-modulated transmission. SSB communication is sometimes chosen for short-distance operation when a number of channels are required. When few channels are required, a frequency-modulated (FM) transmitter may be employed. The chief advantage of FM transmission is a very high signal-to-noise ratio, but FM transmission is comparatively wasteful of the available frequency spectrum. FM transmitters used in power-line carrier systems provide an improvement in signal-to-noise ratio of 4.8 dB compared with a conventional AM transmitter. The deviation ratio of a power-line carrier FM transmitter is 1 to 1. Better signal-to-noise ratios could be realized by using higher deviation ratios. Commercial FM broadcast stations employ a 1 to 5 deviation ratio. However, the deviation ratio that can be utilized in a power-line carrier system is limited by the established 6000-Hz channels.

Repeaters cannot be used in power-line carrier systems because of high voltages and currents in the lines. Therefore, comparatively-high carrier power must be supplied by the transmitter for long-distance communication. A typical system supplies a carrier output power of 350 watts into the transmission line. Although this is a comparatively-high carrier power level, the distance that can be covered could be rather limited by the characteristics of the transmission line. The most serious limitation is imposed by branch lines tapped into the main transmission line. A typical line without any tap lines produces a carrier loss of 0.13 dB per mile. In turn, a 60-mile line would impose a loss of approximately 8 dB, or a carrier-voltage loss of about 60 percent. Mismatches at each station will usually add another decibel of loss. Next, if there is a tap line between the two stations, half of the available carrier energy is diverted down the tap line. In turn, the total carrier loss in this example becomes 12 dB, or a carrier voltage loss of approximately 75 percent. In other words, only 25 percent of the original carrier voltage remains under this condition.

The foregoing example assumes that the tap line is very long so that effectively it has the same characteristic impedance as that of the main line (typically 800 ohms). In many situations, however, a tap line is comparatively short and is terminated by a power transformer that has a very high impedance at the carrier frequency. In turn, most of the carrier energy is reflected back from the transformer. Consequently, standing waves are established on the line, with the result that carrier energy is nearly zero at some points and higher than normal at other points. For instance, with a carrier frequency of 60 kHz, a tap line 3.875 miles long would absorb most of the carrier energy. On the other hand, a tap line 4.65 miles long would absorb very little carrier energy. Again, a tap line 2.352 miles long would absorb most of the carrier energy. On the other hand, a tap line 3.10 miles long would absorb very little carrier energy. It is usually necessary to determine the characteristics of a transmission line experimentally. In most cases, the transformer characteristics at the carrier frequency are more or less unknown, which makes calculation impractical.

8.6   WIRELESS INTERCOM UNITS

Previous mention of wireless intercoms was made in Sec. 2.2. Wireless intercommunication units are simplified power-line carrier transmitters and receivers. Typical wireless intercom units are illustrated in Fig. 8–14. Each unit contains a small loudspeaker that doubles as a microphone during transmission. The intercom units are plugged into 120-volt 60-Hz outlets for a power source and also to couple the carrier into the line. This type of power-line carrier equipment is designed for local operation only at distances up to 100 feet. Therefore, low-power transistors are employed. Each unit is provided with an off-on switch, volume control switch, and press-

Fig. 8–14.   Typical wireless intercom unit. (Courtesy of Heath Co.)

to-talk switch. Voice calling is utilized and the units are quite compact. Since a low-power carrier is used, the signal-to-noise ratio is not as high as in wired intercom systems. However, the background noise is not unduly disturbing and the economy and convenience of a wireless system are attractive features.

Conventional wireless intercom installations are of the party line type, in which all stations can listen-in to a conversation. It is not possible to interrupt a speaker because the receiver is switched off during transmitter operation. It is necessary for a pair of stations to have a continuous line between the plug-in points. In other words, communication cannot be made between adjacent buildings that are serviced by different line transformers. However, if the buildings are serviced by the same line transformer, communication is usually possible. If the length of line is excessive, received signals will be weak and objectionably noisy.

9

NEW-MUSIC
AUDIO SYSTEMS

9.1   GENERAL CONSIDERATIONS

There are various types of new-music devices and systems, all of
which represent departures from traditional musical instruments
and musical forms. One of the most active areas in new music is
termed the *electrophonic method*, which employs electronic oscillators or electrical generators, filters, switching arrangements, and
speakers. A new-music electrophonic arrangement is more or less
related to electronic-organ technology but differs in several important ways. Although an electronic organ is directly under the
control of a performer, an electrophonic instrument is only indi-
rectly controlled by a performer, or, it may employ various digital-
computer techniques to make it self-programming. The latter class
of new-music instruments are designed to replace both the com-
poser and the musician. Electrophonic instruments are sometimes
called mechanical-music sources. They may be actuated by

punched-paper tape, such as employed by various digital computers. However, this type of mechanical-music source cannot be logically compared with a player piano, which is also actuated by a roll of punched-paper tape, because a player piano produces music in the traditional forms, whereas a mechanical-music source produces *new sounds* that are not necessarily related harmoniously nor by any of the established rules of musical composition.

An electrophonic instrument may be of the melodic or polyphonic type; that is, it may play sounds sequentially or it may play *chords*. New music seeks primarily to be *different*. Sounds that are usually classified as noise are often utilized with various special effects. As an illustration, the *fuzz box*, a teen-age musical distortion device, and the *wow-wow* box, are often included in new-music systems. As explained previously, wow-wow denotes a very slow vibrato. Numerous other distorting devices are also used, as detailed subsequently. Random noise, such as produced by a TV receiver on a vacant channel, may be introduced directly or modified in some way. A familiar sound or series of sounds may be reversed in sequence by means of magnetic tape that is played back from end to beginning. Or the tape may be cut into a number of sections that are spliced together in a random manner. The new art is in a state of ferment and much experimentation is in progress.

9.2   MUSICAL TONE PARAMETERS

Musical tones occupy a wide spectrum that grade imperceptibly from traditional structures into various nonmusical sounds and noise. Traditional musical tones are characterized by pitch (frequency), loudness (amplitude), and timbre (harmonic content). These are steady-state parameters that describe the characteristics of a note that is produced by holding down a key of an electronic organ, for example. In addition to the steady-state parameters, a musical tone is also characterized by certain transient parameters. Thus, musical tones are described also in terms of attack (rise mode and rise time), duration (sustain time), and decay (fall mode and fall time). The attack of a tone is sometimes termed its *growth*. Musical tones may also be characterized by a tremolo (amplitude-modulated quality), vibrato (frequency-modulated quality), or a portamento (frequency glide) quality. As noted previously, tremolo and vibrato may be combined; moreover, either or both may be used with a frequency glide.

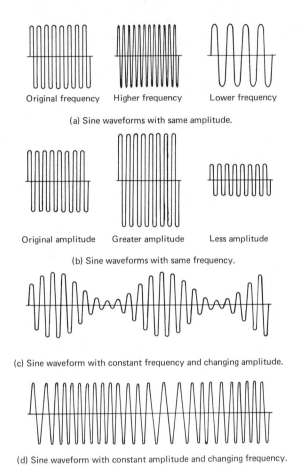

(a) Sine waveforms with same amplitude.

(b) Sine waveforms with same frequency.

(c) Sine waveform with constant frequency and changing amplitude.

(d) Sine waveform with constant amplitude and changing frequency.

Fig. 9–1. Basic sine-wave frequency and amplitude characteristics.

Three different pitches, or frequencies, are depicted in Fig. 9–1(a) as they would appear on the screen of an oscilloscope. Three different degrees of amplitude, or loudness, are represented in Fig. 9–1(b). Tremolo is exemplified in Fig. 9–1(c) and vibrato is depicted in Fig. 9–1(d). The timbre of six musical tones is illustrated in Fig. 9–2. Note that a tuning fork has nearly a pure sine waveform. Figure 9–3 shows attack, sustain, and decay periods. It is an example of symmetrical attack and decay intervals. However, in many cases the decay time is much shorter or much longer than the attack time. A rapid frequency glide, or portamento, is shown

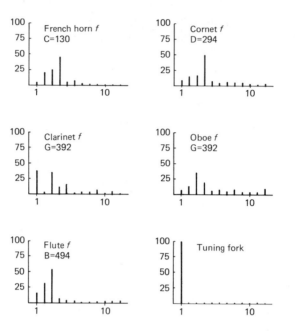

Fig. 9–2. Frequency spectra (timbre) of six musical tones.

in Fig. 9–4. Note that the phase relations among the fundamental and harmonics of a musical tone are of no significance with respect to its audio characteristics. For example, Fig. 9–5 depicts a fundamental and its third harmonic in six different phase relations. Insofar as the human ear is concerned, all six waveforms correspond to exactly the same tone. This is one of the factors that makes it difficult to predict the subjective effect of an arbitrary complex waveform.

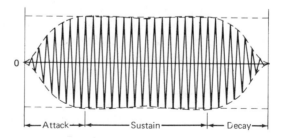

Fig. 9–3. Example of attack, sustain, and decay.

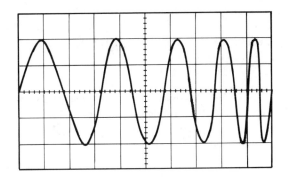

Fig. 9–4. Example of a rapid frequency glide.

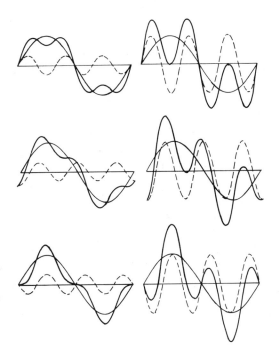

Fig. 9–5. Waveforms produced by a fundamental and its third harmonic in various phase relations.

9.3 BASIC NEW-MUSIC SOUND PARAMETERS

In addition to the sine wave, new-music systems make extensive use of square, sawtooth, and triangular waveforms, as shown in Fig. 9–6. A sine wave corresponds to a pure and thin sound whereas a square wave has a comparatively harsh timbre. A rectangular waveform, such as depicted in Fig. 9–7, is essentially an unsymmetrical square wave. The rectangular waveform contains both even and odd harmonics whereas a square wave contains only odd harmonics. Because of its even-harmonic content, a rectangular waveform has a fuller timbre than a square wave. A sawtooth wave also has a fuller timbre than a square wave owing to its even-harmonic content. On the other hand, a triangular waveform, which has odd harmonics only, is more nearly comparable to a square wave. However, since these odd harmonics are comparatively low in amplitude, the timbre of a triangular wave is intermediate to that of a square wave and a sine wave. It is instructive to note the relation of a sawtooth frequency spectrum to standard musical pitches, as shown in Fig. 9–8.

Note that a square wave (Fig. 9–6) has an extremely rapid attack whereas a triangular waveform has a slow attack. A rapid attack is associated with higher amplitude harmonics. Many more of the harmonics must be reproduced to maintain a very rapid attack. Measurement of the attack period is made as shown in Fig. 9–9. The attack period is defined as the rise time of wavefront. It is the time required for the voltage to rise from 10 percent to 90 percent of its final value. Figure 9–10 illustrates the manner in which the fundamental and harmonics of a square wave combine to form the complex waveshape. It follows from the three examples that the rise time of the complex waveform will be slowed if the higher harmonics are removed. This is just another way of saying that a rapid attack cannot be maintained unless the waveform is processed by circuits that have ample bandwidth.

The rise time of a square wave can be reduced to any desired extent by passing the original waveform through a low-pass filter. Figure 9–11 shows three low-pass filters of the RC type and their effect on the rise time of a square wave. An RC low-pass filter is often called an *integrator*. Note that when a single RC section is employed, the leading edge of the output waveform attains 63

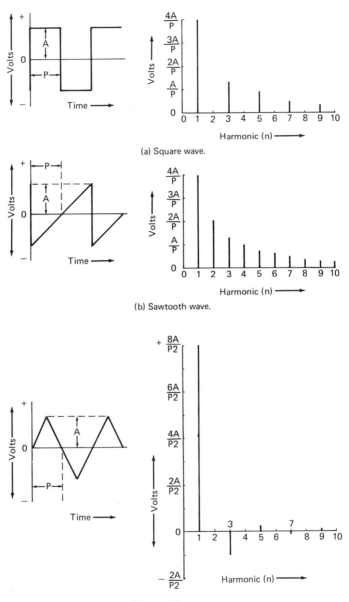

(a) Square wave.

(b) Sawtooth wave.

(c) Triangular wave.

Fig. 9–6.   Square, sawtooth, and triangular waveforms, with their first nine harmonics.

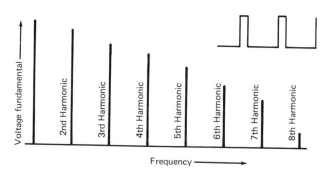

Fig. 9–7. A rectangular waveform, with its first seven harmonics.

percent of its final amplitude in one time-constant. The time-constant of the RC section is equal to the product of the resistance and the capacitance components in the section. If R is expressed in ohms, and C in farads, the time-constant is given in seconds (or a fraction of a section, as the case may be). When two identical RC sections are utilized in an integrator, the output waveform does not attain 63 percent of maximum amplitude until almost three time-constants. Or, if three RC sections are used, the output waveform does not attain 63 percent of maximum amplitude until more than seven time-constants have elapsed.

Next, consider the decay period of the output waveform from a single-section RC integrator when energized by a square wave. As depicted in Fig. 9–12, the decay interval is symmetrical with respect to the growth interval. In other words, the fall time is the same as the rise time when one of the curves is *turned upside down.* This same principle applies to two-section and three-section RC integrators. Thus, with reference to Fig. 9–11, all three curves may be turned upside down to indicate the fall times of the output waveforms. A sine-wave note may have a fast rise and a slow decay to produce a percussive note, as exemplified in Fig. 9–13. The rise is very rapid in this example, and the decay time is prolonged. Note that the time-constant of the wave envelope extends over ten cycles of the sine-wave tone. This type of waveform is usually generated by shock-exciting a ringing circuit with a pulse waveform, as depicted in Fig. 9–14.

When an LC circuit is shock-excited, it oscillates at its natural resonant frequency and generates an exponentially damped waveform. Exponential waveforms are as basic as the sine wave; Fig.

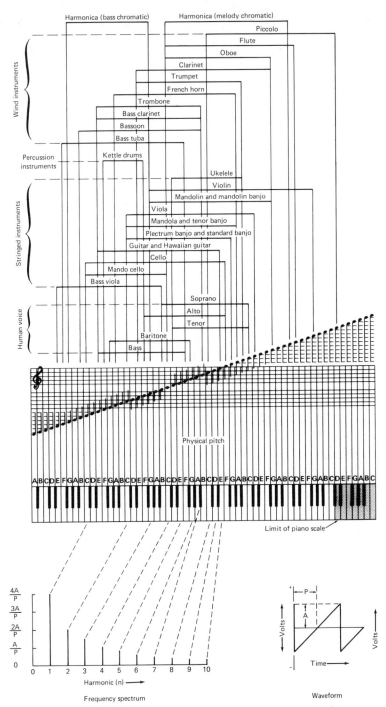

Fig. 9–8. Relations of sawtooth harmonic frequencies to standard musical pitches.

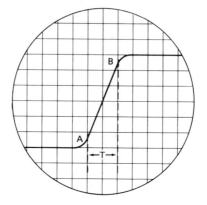

Fig. 9–9.   Rise time of the wavefront is measured from
A to *B*.

9–12 shows examples of exponential waveforms, in addition to the
wave envelope in **Fig. 9–13**. Note that the rate of damping depends
upon the winding resistance of the inductor in the shock-excited
circuit. If the inductor has comparatively little resistance, its *Q*
value (ratio of reactance to resistance) is high, and the circuit will

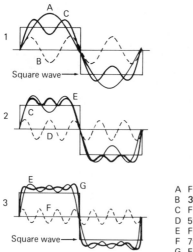

A Fundamental
B 3rd harmonic
C Fundamental plus 3rd harmonic
D 5th harmonic
E Fundamental plus 3rd and 5th harmonics
F 7th harmonic
G Fundamental plus 3rd, 5th, and 7th harmonics

Fig. 9–10.   Partial synthesis of a square wave from its
sine-wave components.

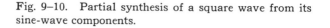

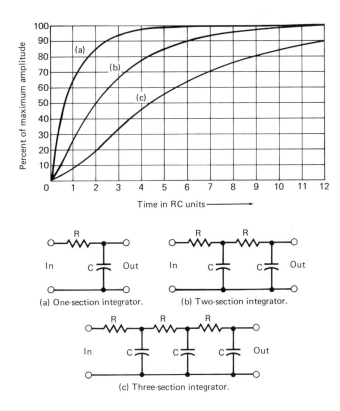

Fig. 9–11.   RC low-pass filters and their effect on the rise time of a square wavefront.

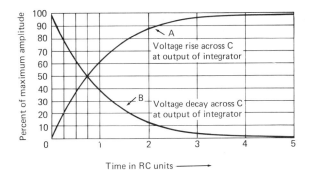

Fig. 9–12.   An RC low-pass filter produces symmetrical attack and decay intervals in its processing of a square wave.

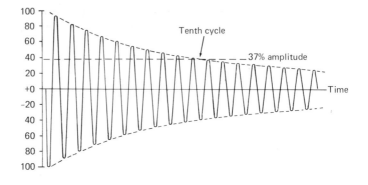

Fig. 9–13. Example of a note with fast rise and slow decay.

ring for a long time. On the other hand, if the inductor has considerable resistance, its *Q* value is low and the circuit will ring for only a short time. If the resistance is sufficiently great that ringing is just barely suppressed, the circuit is said to be critically damped and its response is a single-surge waveshape, as shown in Fig. 9–15. It is instructive to note that a piano tone has an envelope waveform that approximates a critically-damped surge.

Two sine waves that have different frequencies can be combined in two basic ways, as shown in Fig. 9–16. If waveform *A* is mixed with waveform *C* in a linear circuit, the output waveform appears as depicted in *E*. This waveform has the two original frequency components and the ear perceives the waveform as two frequencies or pitches. On the other hand, if waveform *A* is combined with waveform *C* in a nonlinear (modulator) circuit, the output waveform appears as shown in *G*. Now, the output waveform contains a new pair of frequencies and the ear perceives essentially the three frequencies noted in the spectrum for *G*. Because of the non-

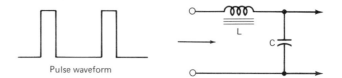

Fig. 9–14. A shock-excited LC circuit generates an exponentially damped sine waveform.

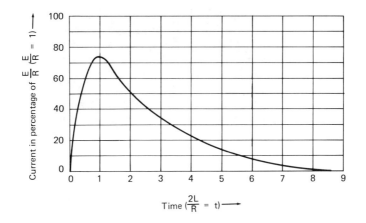

Fig. 9–15.   Surge waveform output from a shock-excited critically damped LC circuit.

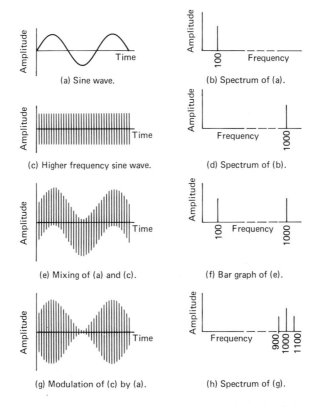

Fig. 9–16.   Linear and nonlinear mixing of a pair of sine waves.

Fig. 9–17. Damped sine waveform that has a reverberation or echo characteristic.

linearity of ear response, frequency *A* may be detectable in some
cases. Finally, if an exponentially-damped sine wave is modulated
as shown in Fig. 9–17, a reverberation or echo effect is perceived.

An audio tone may also be modulated by the carrier-suppres-
sion method, resulting in a timbre distinctively different from that
produced by a conventional amplitude-modulated tone. A balanced

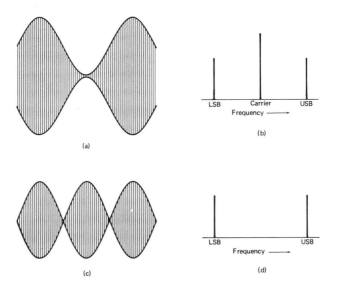

Fig. 9–18. Example of carrier suppression. (a) Carrier
and both sidebands present in waveform. (b) Components of waveform in (a). (c) Both sidebands without
carrier in waveform. (d) Components of waveform
in (c).

modulator may be employed for carrier suppression, as explained previously. The result of carrier suppression is the production of a double-frequency wave envelope, as depicted in Fig. 9–18. The modulating frequency also feeds through (see Fig. 9–19) unless it is trapped by a filter in the output circuit of the modulator. Thus, the output is a mixed waveform. If modulating-frequency feedthrough is not desired, a lattice modulator may be employed as an alternative to trapping the modulating frequency in the output circuit. With reference to Fig. 9–20, a lattice modulator suppresses both the modulating frequency and the modulated frequency, leaving only the two sideband frequencies. In turn, a lattice modulator has general utility in system design inasmuch as a trap is effective at only one frequency.

Modification of an audio tone is also effected by rectification,

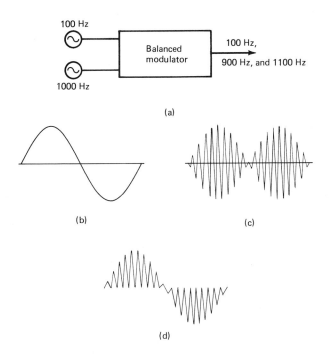

(a)

(b)

(c)

(d)

Fig. 9–19. Example of tone modulation by a balanced modulator. (a) Modulator frequency relations. (b) Modulating waveform. (c) Sidebands with suppressed carrier. (d) Mixed waveform—sidebands plus modulating waveform.

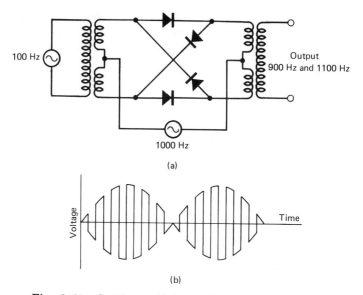

(a)

(b)

Fig. 9–20. Lattice modulator eliminates modulating frequency feedthrough. (a) Modulator configuration. (b) Suppressed-carrier output waveform.

as exemplified in **Fig. 9–21.** For example, a 100-Hz sine wave produces a smooth humming sound while a rectified sine wave produces a buzzing sound. Rectification is a special case of clipping. For instance, if a transistor rectifier circuit employs signal-developed bias, as depicted in **Fig. 9–22,** negative-peak clipping results. This output waveform has a distinctly different timbre than that of a half-rectified waveform. Again, the arrangement shown in **Fig.**

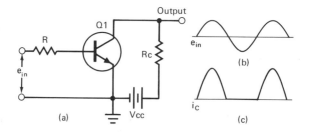

Fig. 9–21. Half rectification of a sine waveform. (a) Transistor rectifier configuration. (b) Input waveform. (c) Output waveform.

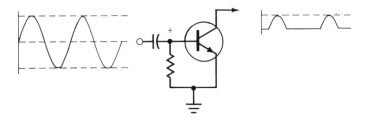

Fig. 9–22.   Example of a negative-peak clipper.

9–23 can be used to slice off both the positive and negative peaks from an input waveform. This configuration utilizes zener diodes that conduct at levels E_{z1} and E_{z2}. In turn, the peaks of the input waveform are shunted to ground and do not appear in the output circuit. A circuit that provides an output up to a specified voltage is also called a limiter.

9.4   ORGANIZATION OF A NEW-MUSIC SYNTHESIZER

A simple new-music synthesizer system in the Moog tradition is depicted in Fig. 9–24. Its output is entirely predictable inasmuch as the sequence of tones is determined by perforations in a pre-pared paper tape. From eight to twelve tracks are generally pro-vided on the punched tape. Light beams and photoelectric devices are typically utilized to sense the passage of a perforation in the tape. In turn, the instructions from the tape activate electronic switches in a switching unit. Each switch will turn a tone generator off or on, as in the example of Fig. 9–24. Thus, the sine-wave, square-wave, or sawtooth generator can be switched on either in-

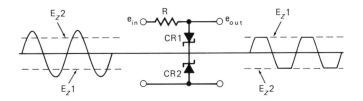

Fig. 9–23.   A Zener-diode dual peak clipper.

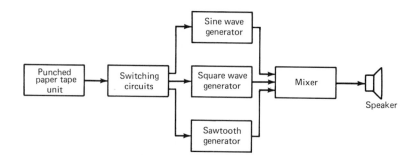

Fig. 9–24.   A simple new-music synthesizer system.

dividually or simultaneously with the others. Outputs from the tone generators are fed into a linear mixer and then into a speaker. Synthesizer units are typically constructed in module form and interconnected with patch cords. This arrangement provides considerable flexibility in operation. As an illustration, an additional triangular waveform generator could be patched between the switching circuits and the mixer.

Another basic synthesizer system employs a random noise generator instead of a punched paper-tape unit. The noise generator makes the output from the speaker unpredictable since random noise has no definite pattern. A high-gain audio amplifier has a comparatively high-level noise output. The noise waveform consists of a series of pulses that follow no orderly sequence. If the amplifier has a wide bandpass, noise pulses are produced rapidly (many pulses per second). On the other hand, if the amplifier has a narrow bandpass, noise pulses are produced slowly (few pulses per second). Noise pulses will operate the switching circuits in Fig. 9–24 in the same basic manner as pulses from photoelectric devices. Note that noise pulses have random amplitudes as well as random occurrence in time. Therefore, not all of the random noise pulses will have sufficient amplitude to actuate the switching circuits. In application, the switching circuits are biased to different thresholds. Consequently, low-amplitude pulses will operate one switch, medium-amplitude pulses will operate another switch, and high-amplitude pulses will operate still another switch. Very low-amplitude pulses will not operate any of the switches.

Figure 9–25 shows a synthesizer system that employs both predictable and unpredictable operating factors. The punched paper-tape unit controls the sine-wave generator and the sawtooth generator; outputs from these generators are predictable. On the

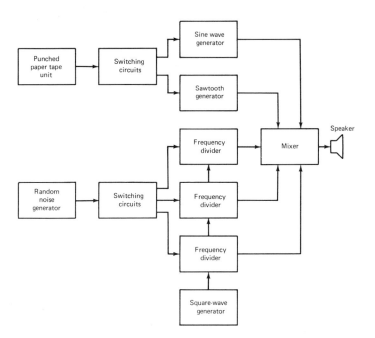

Fig. 9–25. A synthesizer system that employs both predictable and unpredictable operating factors.

other hand, the random-noise generator controls three frequency
dividers which are energized by a square-wave generator. Therefore, the square-wave outputs are unpredictable. Both the predictable and unpredictable outputs are mixed together and fed to the
speaker. Note that the frequency-divider units are designed as
bistable multivibrators for division-by-two, as explained previously.
Thus, the unpredictable output is always a square wave but its
frequency may have any one of three values, two of these values
simultaneously, or all three values simultaneously. An elaboration
of this arrangement includes Schmitt trigger circuits that are
biased to produce rectangular wave outputs instead of square-
wave outputs. If the bias voltage is fixed, the output is a predictable rectangular waveshape. On the other hand, if the bias voltage
is determined by amplitudes of random noise pulses, the output
waveshape (proportions of long and short intervals) is unpredictable. Thus, the duration of a tone can be controlled in addition to
its frequency. Another elaboration provides predictable or unpredictable control of loudness.

Figure 9–26 shows the configuration of a basic Schmitt trigger

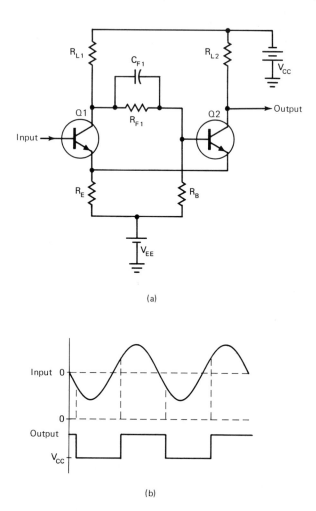

Fig. 9–26. A basic Schmitt trigger circuit. (a) Config-uration. (b) Triggering sequence.

circuit. It changes state at a certain negative-voltage level and then changes its state back at a certain positive voltage level. The output is always a rectangular waveform, regardless of the input trigger waveform. Although a sine-wave input is depicted in Fig. 9–26, a pulse-voltage input as exemplified in Fig. 9–27 is often employed in synthesizer systems. The spacing of consecutive pulses deter-mines the duty cycle (off versus on intervals) of the rectangular waveform output. As noted previously, the pulse input may be

Fig. 9–27. Example of pulse sequence from a digital-computer punched tape.

predetermined (as from a punched tape) or it may be indeterminate (as from random noise pulses). If desired, a Schmitt trigger circuit may be followed by a chain of divide-by-two bistable multivibrators to obtain a series of octavely related tones. Diode or transistor switches may be utilized to turn AM or FM modulators on and off. Various types of formant filters may be used as explained previously to modify the timbre of any tone.

More sophisticated types of switching arrangements utilize digital-computer gating circuitry, as exemplified in Fig. 9–28. An OR gate produces an output if either of two input pulses is applied. On the other hand, an AND gate produces an output only if both of two input pulses are applied simultaneously. A **NOR** gate is

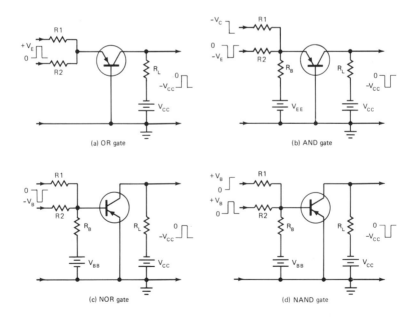

Fig. 9–28. Computer-type switching gate configurations.

essentially the same as an OR gate, except that the polarity of the input pulse is inverted at the output. A NOT-AND gate is essentially the same as an AND gate, except that the polarity of the input pulses is inverted at the output. Gate circuits may employ diodes instead of transistors; however a transistor has an advantage over a diode in that it provides amplification in addition to switching action. It is evident that many types of new-music synthesizer systems can be devised, ranging from very simple assemblies to extremely complex assemblies. This new art has not yet crystallized into established forms. However, the foregoing discussion is descriptive of the most prominent present-day trend.

10

AUDIO
MEASUREMENTS

10.1   GENERAL CONSIDERATIONS

Audio measurements include basic values such as voltage, current, and power. Resistance measurements may be made to determine a resistance value or to calculate a current value from the voltage drop across the resistor. Frequency is often measured to determine the uniformity of amplifier response and to check the low and high frequency cut-off points. In addition, distortion is measured in terms of either percentage harmonic distortion or percentage intermodulation distortion. Transient response may be checked in terms of square-wave distortion. Amplifier voltage or power gain is measured in terms on the input/output voltage ratio or power ratio. Noise is measured in terms of output voltage when no input voltage is applied to an amplifier or system. Hum is usually measured by an oscilloscope because the operator can distinguish between a hum pattern and a random noise pattern in the screen display. Separation is measured in stereo systems on the basis of dB values. Various other audio measurements can also be made, although these are the most important.

10.2 CHARACTERISTICS OF A SINE WAVE

A sine wave can be described in terms of its root-mean-square (rms) value, peak value, or peak-to-peak value, as exemplified in Fig. 10–1. Note that the rms value of a sine wave is equal to 70.7 percent of its peak value. These values apply to sine waves of voltage, current, or power. A wide range of voltages is encountered in audio systems. Therefore, the audio technician has need of a wide-range audio voltmeter, such as illustrated in Fig. 10–2. Although a conventional transistor voltmeter (TVM) has considerable usefulness in audio measurements, this type of instrument is limited in its range of low-voltage response. For example, a low-level amplifier may process a signal at a level of a few millivolts. In turn, a TVM will provide no useful response; only an audio voltmeter can be used in this type of application. Most audio voltmeters indicate the rms value of a sine wave.

A sine wave has equal positive-peak and negative-peak values, provided that it is undistorted. On the other hand, a distorted sine

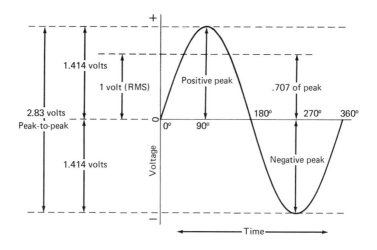

Fig. 10–1. Basic values of a sine wave.

Fig. 10–2. A typical audio voltmeter.

wave may have unequal positive-peak and negative-peak values, as seen in Fig. 10–3. Small percentages of distortion are difficult or impossible to evaluate either by eye or by ear. As explained in greater detail shortly, special instruments are required to measure distortion percentages accurately. Note that one class of distorted sine wave has equal positive-peak and negative-peak voltages. Therefore, it is not feasible to attempt to evaluate distortion on the basis of peak-voltage measurements.

10.3 POWER OUTPUT MEASUREMENT

Power output of an amplifier is generally measured as depicted in Fig. 10–4. The amplifier is energized by an audio oscillator and is loaded by a power resistor of suitable rating. Note that R should have a resistance value equal to the rated output impedance of the amplifier and should be capable of dissipating the maximum-rated power output of the amplifier. The power value is measured in terms of the voltage drop across the load resistor, in accordance with the power law: $P = E^2/R$. The audio oscillator should provide

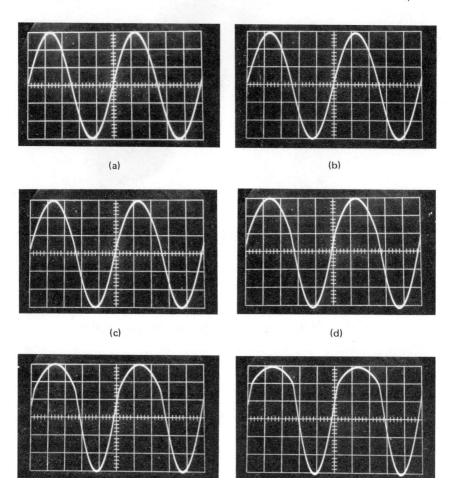

Fig. 10–3. Sine waves with various percentages of distortion. (a) 1%. (b) 3%. (c) 5%. (d) 10%. (e) 15%. (f) 20%.

a good sine waveform, and the amplifier must not distort the sine wave appreciably if the measurement is to be accurate. In other words, most AC voltmeters indicate incorrect voltage values in case of waveform error. Although a true rms-reading voltmeter can be employed, this type of instrument is comparatively expensive. An oscilloscope can be connected across the output of the amplifier to indicate serious distortion (see Fig. 10–3).

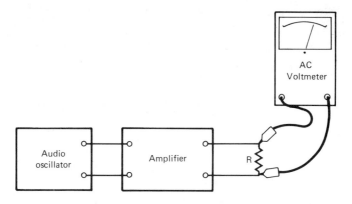

Fig. 10–4.   Measurement of amplifier power output by
calculation from rms output voltage.

10.4   FREQUENCY RESPONSE
MEASUREMENTS

The frequency response of an amplifier can be measured with the
same test setup depicted in Fig. 10–4. It is advisable to check frequency response at maximum rated power output since the frequency response of some amplifiers becomes less as the power output is increased. Typical audio amplifier response curves are shown
in Fig. 10–5. Note that the frequency response cannot be checked
accurately unless the bass and treble controls are set to mid-range,
the loudness control set for flat response, and the scratch and
rumble filters turned off. Frequency response tests are greatly facilitated by the use of an audio oscillator that has uniform output
over its entire frequency range. Otherwise, the output voltage of the
audio oscillator must be monitored with a meter so that variations
in the input signal level will not be falsely charged to amplifier
characteristics. Note that the voltage gain of an amplifier at any
frequency is equal to the ratio of its output voltage to input voltage.

10.5   SQUARE-WAVE TESTING

Square-wave tests are informative because they show the transient
response of an amplifier whereas a frequency response check shows
the steady state response of the amplifier. Although transient and

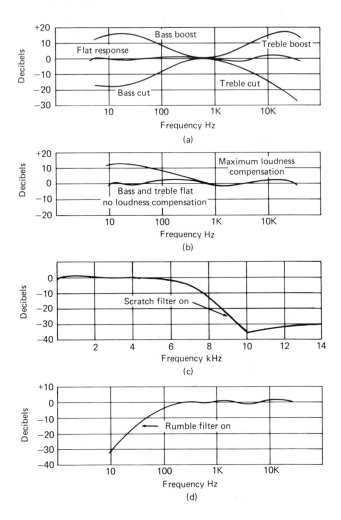

Fig. 10–5. Typical amplifier frequency response curves. (a) Flat response with effects of bass and treble controls. (b) Flat response with effect of loudness control. (c) Frequency response of a scratch filter. (d) Frequency response of a rumble filter.

steady state responses are related, it is difficult to calculate the transient response from frequency response data. A complete calculation involves the phase characteristic of the amplifier. Therefore, it is much simpler to check transient response by means of a square-wave test. Figure 10–6 shows the test setup that is utilized

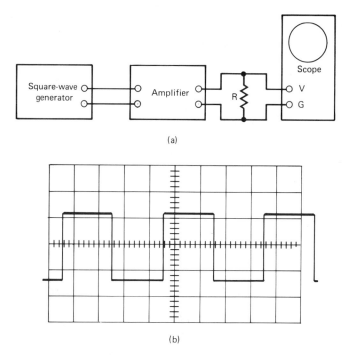

(a)

(b)

Fig. 10–6. Square-wave testing of an audio amplifier. (a) Test setup. (b) Typical 2 kHz square-wave response of a hi-fi amplifier. (Courtesy of General Electric Co.)

and an example of the 2-kHz square-wave response for a high-fidelity amplifier. Manufacturers sometimes rate hi-fi amplifiers for square-wave response. A valid test requires that the square-wave generator have a faster rise time than the amplifier to be tested; it is also necessary that the oscilloscope have better square-wave response than the amplifier. In general, laboratory-type instruments are advised for meaningful square-wave tests. Figure 10–7 shows typical distortions of square waves, caused by various amplifier defects. If an amplifier distorts a square wave, it will also distort musical waveforms.

It is advisable to make a square-wave test at maximum rated power output of the amplifier. In other words, the square-wave response at maximum power level may not be as good as the response at a low power level. The bass and treble controls should be set to mid-range, the loudness control set for flat response, and the scratch and rumble filters turned off. Otherwise, the square-wave

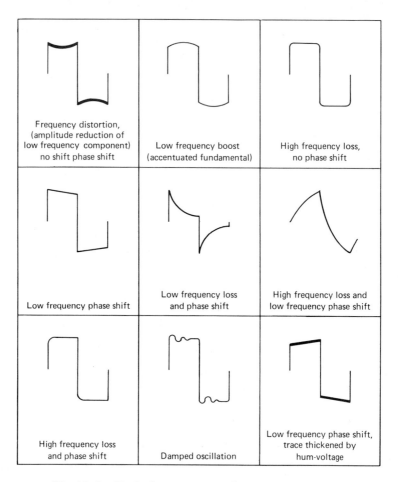

Fig. 10–7.   Typical square-wave distortions.

response will be affected and the distortion will be falsely charged to amplifier characteristics. Although it is standard practice to check hi-fi amplifiers with a 2-kHz square wave, it is also informative to check the response with a 60-Hz square wave. If the top of the reproduced square wave is not tilted, it is indicated that the amplifier provides very good low-frequency transient response. Note also that any hum voltage will cause the pattern to *writhe*. In other words, if the hum voltage has a frequency of 60 Hz, and the square wave has a frequency of 61 Hz, the pattern will writhe at a 1-Hz rate. Again, if the hum voltage has a frequency of 120 Hz, and the square wave has a frequency of 119 Hz, the pattern will writhe at a 1-Hz rate.

10.6   PERCENTAGE DISTORTION MEASUREMENTS

Harmonic distortion is measured with a test setup such as shown in Fig. 10–8. The audio oscillator must have an output that has lower distortion than the amplifier under test. It is standard practice to measure percentage harmonic distortion at 1 kHz. However, distortion can be checked at higher frequencies and at lower frequencies for more comprehensive test data. A harmonic distortion meter contains a tuneable filter which is adjusted by the operator to eliminate the fundamental frequency from the amplifier output signal. In turn, only the harmonics generated by amplifier distortion remain to energize the meter. The meter scale is read directly

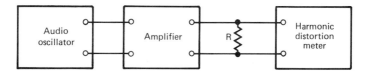

(a)

(b)

Fig. 10–8.   Measurement of percentage harmonic distortion. (a) Test setup. (b) Appearance of harmonic distortion meter. (Courtesy of Heath Co.)

in terms of percentage harmonic distortion. Before the reading is taken, the input signal level to the harmonic distortion meter is adjusted to a reference level. This adjustment is required to make the meter scale direct-reading. If desired, an oscilloscope can be connected at the output of the harmonic distortion meter to display the distortion. In this manner the operator can check for even and odd harmonics and for presence of hum voltage.

10.7   PERCENTAGE INTERMODULATION MEASUREMENTS

Percentage intermodulation distortion is measured with a test setup such as shown in Fig. 10–9. This is a two-tone test that employs

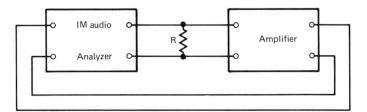

(a)

(b)

Fig. 10–9.   Measurement of percentage intermodulation distortion. (a) Test setup. (b) Appearance of intermodulation audio analyzer. (Courtesy of Heath Co.)

signals generated by the analyzer; frequencies of 60 Hz and 6 kHz are commonly utilized. If the amplifier operates in a linear manner, these two frequencies are unaffected by passage through the amplifier. On the other hand, if the amplifier operates in a nonlinear manner, the 60-Hz signal modulates the 6-kHz signal more or less. In turn, a beat or difference frequency is produced which appears at the output of a rectifier, as depicted in Fig. 10–10. Low-pass and high-pass filters are provided to eliminate the test signals and to leave only the modulation envelope of the signal to energize the

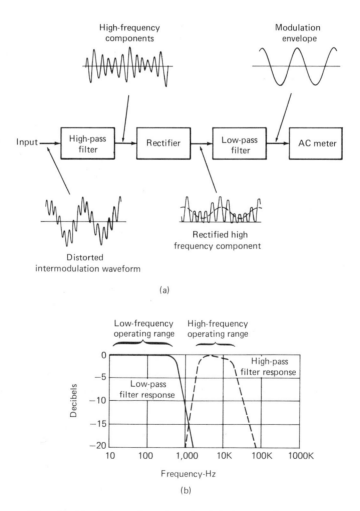

Fig. 10–10. Plan of an intermodulation analyzer. (a) Block diagram. (b) Characteristics of filters.

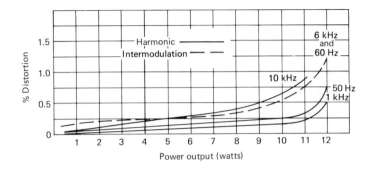

Fig. 10–11. Example of harmonic and intermodulation distortion percentages for a high-fidelity amplifier. (Courtesy of General Electric Co.)

meter. The meter scale is direct-reading in terms of percentage intermodulation distortion. In general, harmonic distortion and intermodulation distortion tend to be consistent. However, there are some differences encountered over the audio frequency and power ranges, as exemplified in Fig. 10–11.

10.8 PHASE-SHIFT MEASUREMENTS

All amplifiers have zero phase shift between input and output signals at some frequency in the vicinity of their mid-frequency range. Negative-feedback loops modify the phase characteristic of an amplifier. In general, the phase shift becomes rapid near the low-frequency and high-frequency cut-off points of the amplifier. Figure 10–12 shows a test setup for measurement of phase shift at any frequency over the audio range with examples of phase-shift patterns. These Lissajous figures go through cyclic changes; however, this is not a source of confusion in amplifier tests because the maximum phase shift that can be encountered is ±90°. Note that if the amplifier distorts the test signal, true ellipses or circles and straight diagonal lines will not be displayed, as depicted in Fig. 10–13. Unless the oscilloscope has linear amplifiers, the deficiencies of the oscilloscope will be falsely charged to the audio-amplifier characteristics. It is also necessary that the audio oscillator provide a good sine waveform to avoid distortion of the Lissajous figures.

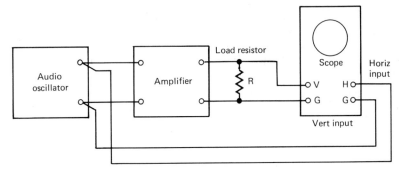

(a)

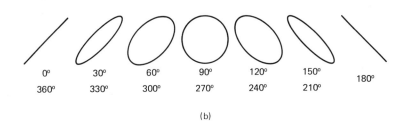

0°	30°	60°	90°	120°	150°	
360°	330°	300°	270°	240°	210°	180°

(b)

Fig. 10–12.   Measurement of amplifier phase shift. (a)
Test setup. (b) Lissajous patterns in 30° increments.

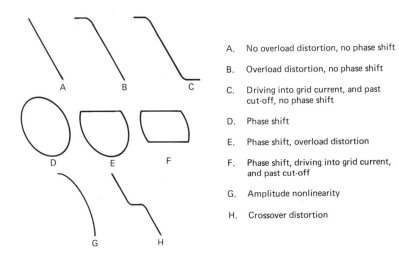

A.   No overload distortion, no phase shift

B.   Overload distortion, no phase shift

C.   Driving into grid current, and past
      cut-off, no phase shift

D.   Phase shift

E.   Phase shift, overload distortion

F.   Phase shift, driving into grid current,
      and past cut-off

G.   Amplitude nonlinearity

H.   Crossover distortion

Fig. 10–13.   Examples of distorted Lissajous figures,
with corresponding amplifier malfunctions.

243

10.9   MULTIPLEX DECODER
SEPARATION TEST

Multiplex decoder separation tests are made with a signal from a stereo analyzer generator, such as illustrated in Fig. 10–14. Indication is obtained with a TVM or scope, as depicted in Fig. 10–15. When a decoder is operating normally, and is energized by a left signal, there will be maximum output from the *L* channel, and very little output from the *R* channel. If the *L* channel is not at least 30 dB down, the presence of decoder trouble is indicated. Conversely, when the decoder is energized by a right signal, normally there will be maximum output from the *R* channel and very little output from the *L* channel. Note that a decoder can also be energized by a test signal passed through the FM receiver, as depicted in Fig. 10–16. If the test signal is checked with an oscilloscope when it enters the decoder, the waveform appears normally as shown in Fig. 10–16(b). On the other hand, if the baseline of the waveform is bowed or curved, it is indicated that the phase characteristic of the FM receiver is incorrect. The remedy in this situa-

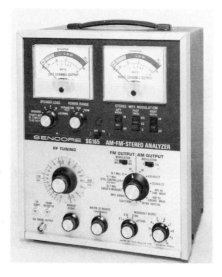

Fig. 10–14.   A stereo analyzer generator. (Courtesy of Sencore)

tion is to realign the circuits of the **FM** tuner in accordance with service-data specifications. Poor frequency response, or incorrect phase characteristics in the **FM** receiver, will cause impaired separation.

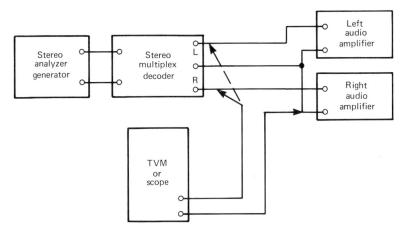

Fig. 10–15.   Multiplex decoder separation test.

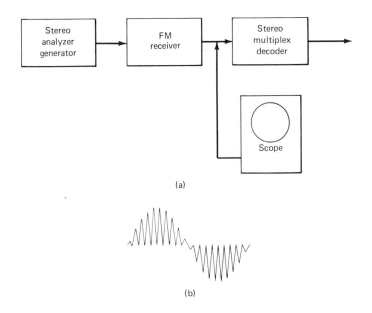

(a)

(b)

Fig. 10–16.   Check of stereo-multiplex waveform through **FM** receiver. (a) Test setup. (b) Normal waveform.

GLOSSARY

A

A-B test—Comparison of sound from two sources, such as comparing original program to tape as it is being recorded by switching rapidly back and forth between them.

Acetate backing—A standard plastic base for magnetic recording tape.

AES—Abbreviation for Audio Engineering Society; also refers to disc-recording curve specified by the AES that is no longer in use.

AF—Abbreviation for audio frequency, a range which extends from 20 to 20,000 Hertz.

AFC—Abbreviation for Automatic Frequency Control, a circuit commonly used in FM tuners to compensate for frequency drift, thus keeping the tuner "locked" onto a station.

AM—Amplitude modulation, method of superimposing intelligence on a radio-carrier signal by varying the amplitude of the carrier.

Amplification—Magnification or enlargement.

Amplifier—An electronic device that magnifies or enlarges electrical voltage or power.

Attenuation—Opposite of amplification, a reduction of electrical voltage or power.

Audio—A term relating to sound.

Audiophile—One who enjoys experimenting with high-fidelity equipment and is likely to seek the best possible reproduction.

B

Background noise—Noise inherent in any electronic system.

Baffle—A barrier separating sound waves generated by the front and back of a speaker cone.

Bass-reflex enclosure—A speaker enclosure with a bass port designed to reverse the phase of a speaker's backwave while using it to reinforce the front wave.

Binaural—A type of sound recording and reproduction. Two microphones, each representing one ear and spaced about 6 inches apart, are used to pick up the material to be recorded on separate tape channels. Playback is accomplished through separate amplifiers (or a two-channel amplifier) or special headphones wired for binaural listening.

C

Capstan—The spindle or shaft of a tape transport mechanism that actually drives the tape past the heads.

Capture ratio—An FM tuner's ability to reject unwanted signals occurring at the same frequency as the desired signal. If an undesired signal, for example, is only 2.2 dB less than a desired one, the undesired signal will be rejected. The lower the capture ratio figure, the better is the tuner.

Cardioid pattern—A heart-shaped (directional) microphone-pickup pattern that tends to reject background noise.

Cartridge—See Pickup cartridge.

Ceramic—A man-made piezo-electric element that is used as the basis of some phonograph pickups; it emits electric current when mechanically strained.

Changer—A record-playing device that automatically accepts and plays up to 10 or 12 records in sequence.

Channel—A complete sound path. A monophonic system has one channel; a stereophonic system has at least two full channels. Monophonic material may be played through a stereo system with both channels carrying the same signal. Stereo material played on a monophonic system mixes and emerges as a monophonic sound. An amplifier may have several input channels, such as microphone, tuner, and phonograph.

Channel balance—Equal response from left and right channels of a stereo amplifier. A balance control on stereo amplifiers permits adjustment for uniform sound volume from both speakers of a hi-fi system.

Chassis—The metal frame, or box, that houses the circuitry of an electronic device.

Compensator—The fixed or variable circuit built into a preamplifier that compensates for bass and treble alterations that were made during the recording process.

Compliance—Physical freedom from rigidity that permits a stylus to track a record groove exactly or a speaker to respond to the audio signal precisely.

cps—Abbreviation for cycles per second. *See* Hertz, Cycle, and Cycles per second.

Crossover network—A filtering circuit that selects and passes certain ranges of audio frequencies to the speakers designed to reproduce them.

Crosstalk—In stereo high fidelity equipment the amount of left channel signal that leaks into the right channel, and vice versa.

Crystal—A natural piezo-electric element used in some phonograph pickup cartridges and microphones; it emits electric current when mechanically strained.

Cycle—One complete reversal of an alternating current, including a rise to a maximum in one direction, a return to zero, a rise to a maximum in the other direction, and another return to zero. The number of cycles occurring in one second is the frequency of an alternating current. The word *cycle* is commonly interpreted to mean cycles per second, in which case it is a measure of frequency. The preferred term is Hertz.

Cycles per second (*cps*)—An absolute unit for measuring the frequency or "pitch" of sound, various forms of electromagnetic radiation, and alternating electric current. *See* Hertz.

D

Damping—The prevention of vibrations, response, or resonances which would cause distortion if unchecked. Control is usually by friction or resistance.

Decibel (dB)—The unit for measuring the intensity or volume of sound; 0 dB is the threshold of human hearing, and 130 dB is the threshold of pain.

De-emphasis—The attenuation of certain frequencies; in playback equalization this offsets the pre-emphasis given to the high frequencies during recording.

Diaphragm—A thin, flexible sheet which vibrates when struck by sound waves, as in a microphone, or which produces sound waves when moved back and forth at an audio rate, as in a headphone or loudspeaker.

Distortion—The deviations from the original sound that crop up in reproduction. Harmonic distortion disturbs the original relationship between a tone and other tones naturally related to it. Intermodulation distortion (IM) introduces new tones caused by mixing two or more original tones.

Ducted port—A cousin of the bass-reflex speaker enclosure in which a tube is placed behind the reflex port.

Dynamic cartridge (electrodynamic)—A magnetic phonograph pickup in which a moving coil in a magnetic field generates the varying voltages of an audio signal.

Dynamic microphone—A microphone that works on basically the same principle as a dynamic cartridge.

Dynamic range—The range of loudness, or sound intensity, that an audio instrument can reproduce without distortion.

E

Effective current—The value of alternating or varying current which produces the same heating effect as the same value of direct current. Also called the rms current.

Efficiency—In a loudspeaker, the ratio of power output to the power input, expressed in percentages—the higher, the better.

Electromagnetic—Pertaining to radiated energy including radio

waves and light waves; a phenomenon involving the inter-
action of electricity and magnetism.

Electrostatic speaker—A type of speaker in which sound is generated by charged plates that are made to vibrate as one is
changed from positive to negative polarity, causing them to
attract and repel each other.

Enclosure—A housing which is acoustically designed for a loudspeaker.

Erase head—The leadoff head of a tape recorder that erases previous recordings from the passing tape by generating a power-
ful random magnetic field.

F

Feed reel—The reel of a tape recorder that supplies the tape.

Fidelity—The faithfulness with which sound is reproduced.

Flat response—The ability of an audio system to reproduce sound
without deviation in intensity for any part of the frequency
range it covers.

Flutter—A form of distortion caused when a tape transport or a
phonograph turntable exhibits rapid speed variations.

FM—*See* Frequency modulation.

FM stereo—Broadcasting over FM frequencies of 2 channels of
sound. Transmitting FM stereo is called *multiplexing*. Stereo
FM (multiplex) tuners are used for FM stereo reception.
Many monophonic FM tuners have been designed to permit
use of an FM stereo adapter.

Folded horn—A type of loudspeaker enclosure using a horn-shaped
passageway which improves the bass response.

Frequency—The number of complete cycles of vibrations per second. Also expressed as Hertz (Hz) per second. Bass tones are
expressed in low frequencies, treble tones in high frequencies.

Frequency modulation—A method of broadcasting which varies or
modulates the frequency of the carrier signal, instead of the
amplitude or strength of the signal, as in amplitude modulation (AM). Most of the static and noises in the radio spectrum
are in the form of AM signals. Advantages of FM are almost
complete freedom from atmospheric and man-made interference as well as little or no interference between stations. FM
is the selected high fidelity medium for broadcasting high
quality program material.

Frequency response—A rating or graph which expresses the manner in which a circuit or device handles the different frequencies falling within its operating range. Thus, the frequency response of a high fidelity amplifier may be specified as being essentially flat or uniform between 20 and 20,000 Hertz.

G

Gain—The degree of amplification a signal receives from an amplifying device.

H

Harmonic distortion—See Distortion.

Head—The electromagnetic device used in magnetic tape recording to convert an audio signal to a magnetic pattern, and vice versa.

Headphones—Small sound reproducers resembling miniature loudspeakers used either singly or in pairs, usually attached to a headband to hold the phones snugly against the ears for private listening. Available in monophonic or stereo design.

Hertz—A unit of frequency equal to one cycle per second (Hz). KHz is the abbreviation for 1000 Hz (1000 cycles, or one kilocycle).

Hill-and-dale—A phonograph reproduction system in which the stylus moves up and down instead of sideways, or laterally.

Horn—A type of speaker in which the mechanical vibrations are coupled to the air by a flaring horn-like passageway instead of a cone.

Hum—Noise generated in an audio device by the power line or current.

I

IHFM (IHF)—Refers to the Institute of High Fidelity Manufacturers, now called the Institute of High Fidelity, Inc. A group of manufacturers who devise and publish standards and ratings for high fidelity equipment.

Image rejection—The effect noticed when one station is heard faintly in the background of another. The tuner's ability to eliminate the background station. Expressed in dB; the higher, the better.

Impedance—Unit of measure, given in ohms, for resistance to an alternating current; it must be matched up between audio units that are connected to each other.

Infinite baffle—Speaker mounting arrangement in which the front and back waves of the cone are totally isolated from each other.

Input—Connection through which an electrical current, or signal, is brought into an electronic device.

Intermodulation distortion (IM)—Two distinct and separate frequencies that are mixed by the amplifier to form the sum and difference of the combined frequencies. Expressed as a percentage; the less, the better. *Also see* Distortion.

J

Jack—Female receptacle for a plug-type connector.

L

Lateral system—System of disc recording in which stylus moves from side to side.

Level indicator—A neon bulb, meter, or cathode-ray ("magic-eye") tube, used to indicate recording level.

Load—The device to which electrical energy is supplied; e.g., a speaker constitutes the load of an amplifier.

Loudness control—Device that boosts treble and particularly bass tones in an amplifier as the volume is reduced; used to compensate for the listener's insensitivity to the extreme ends of the audio range.

M

Magnetic tape—Plastic tape with an iron-oxide coating for magnetic recording.

Manual player—A manual record-playing device with a changer-type motor.

Micro—Prefix meaning one one-millionth.

Milli—Prefix meaning one one-thousandth.

Mixing—Blending two or more signals for special effects.

Monophonic—Recording and reproduction systems in which all program material is on one channel (e.g., as opposed to binaural or stereophonic). Monophonic, sometimes referred to

by the older term monaural, is usually, although not necessarily, associated with a one-speaker system.

Multiplexing—System of broadcasting in which two or more separate channels can be transmitted on one **FM** carrier. Usually
used to denote stereo broadcasting.

N

NAB curve—Tape-recording equalization curve set up by National
Association of Broadcasters; it is accepted as the standard for
commercially recorded tapes.

Network—An electrical circuit.

O

Ohm—Unit of electrical resistance.

Output—Connection through which an electrical current, or signal,
passes out of an electronic device.

P

Patch cord—A shielded cable used to connect one audio device to
another.

Phase—The position at any instance which a periodic wave occupies in its cycle. Any part of a sound wave or signal with
respect to its passage in time. Two devices are in phase when
they provide the same parts of sound or signal simultaneously
and out of phase to the extent that one leads or lags behind the
other.

Phase distortion—Disturbance of the natural timing sequence between a tone and its related overtones. The ear can't detect
phase distortion but it is of consequence in television and test
circuits. Expressed in degrees.

Pickup cartridge—A device used with phonographs to convert
mechanical variations of the record groove into electrical
impulses.

Piezolectric—A crystal or ceramic substance that emits an electrical voltage when it is mechanically strained or twisted; it
is used in phonograph pickups and microphones.

Playback head—The last head on a tape recorder or the only head
on a tape player used to convert the magnetic pattern impressed on a passing tape to an audio signal.

Plug-type connector—A mating connector for a jack.

PM—A permanent magnet that is a basic component of most high fidelity speakers.

Polyester backing—A plastic material used as a base for magnetic recording tape. Developed by DuPont as Mylar, it is extremely strong and resistant to the effects of heat and humidity.

Power amplifier—Amplifier used to drive a speaker.

Power output—The maximum power supplied by an amplifier, expressed in watts.

Preamplifier—Amplifying device used to give an extremely weak signal enough strength to drive a power amplifier.

Pre-emphasis—Exaggeration deliberately introduced into high frequencies during recording for technical reasons.

Print-through—Magnetization of a layer of tape by an adjacent layer, usually most troublesome with thin-based tapes.

Q

Quadriphonic—A system whereby sound picked up by four separate microphones is recorded on separate channels and played back through separate channels driving their own speakers.

Quarter-track recorder—A tape recorder that uses one quarter the width of the tape for each recording; two of the four tracks are used simultaneously for stereo recording.

Quieting—Standard of separation between background noise and program material produced by a radio tuner.

R

Radio receiver—Tuner and amplifier unit on one chassis—also called a *tuner-amp*.

Record head—The second head of a tape recorder; it is used to convert an audio signal to a magnetic pattern on the passing tape.

Record-playback head—Head on a tape recorder performing both record and playback functions.

Recording amplifier—Amplifying circuit of a tape recorder used to prepare an audio signal for input to the record head, and bias current to the erase head.

Reverberation—Persistence of sound after its origin has stopped; deliberately introduced in some audio devices by time-delay and feedback techniques to impart feeling of fullness experienced in concert halls.

RIAA curve—Standard disc-recording curve specified by the Record Industry Association of America.

Rolloff—Another term for de-emphasis, deliberate playback attenuation of high frequencies that had been pre-emphasized during recording.

Rumble—Low-frequency vibration created by an electric motor; particularly prevalent in low-quality phonograph turntable motors.

Rumble filter—A low-frequency filter circuit designed to eliminate rumble.

S

Scratch filter—A high-frequency filter circuit to eliminate scratchy sounds.

Selectivity—The measure of the ability of an electronic device to select a desired signal while rejecting those adjacent to it.

Sensitivity—The minimum input signal required by an electronic device, such as a tuner, to deliver a specified output signal.

Separation—The degree to which one channel is kept from being blended with the other. Expressed in dB; the higher, the better.

Signal-to-noise ratio—The degree expressed in dB, by which program material is separated from background noise.

Soft-suspension speaker—Speaker designed without inherent springiness to use the spring effect of a trapped backwave for restorative force.

Speaker—A device that converts electrical impulses into sound.

Stereophonic sound—A system whereby sound picked up by two separated microphones is recorded on separate channels and played back through separate channels driving their own speakers.

Stroboscopic disc—A cardboard or plastic disc used to check the accuracy of turntable speed.

Stylus—Term for phonograph needle.

Super-tweeter—A speaker capable of reproducing the highest frequencies of the audio range.

T

T pad—A three-element fixed attenuator.

Take-up reel—The reel of a tape recorder that winds the tape after it has passed the heads.

Tape deck—Any tape unit without its own power amplifier and speaker, usually also without a case and designed for custom installation in a high-fidelity system.

Terminal—A connecting point.

Tone arm—The pivoted arm on a phonograph that houses the pickup cartridge.

Tone control—A control, usually part of a resistance-capacitive network, used to alter the frequency response of an amplifier so that the listener can obtain the most pleasing sound.

Tracking—The path of a phonograph pickup stylus in following record grooves.

Transducer—A device that converts electrical energy to mechanical energy or vice versa; e.g., pickup cartridges, speakers, microphones.

Transient response—The ability of a speaker to follow sudden changes of sound level fed from an amplifier.

Transistor—A solid-state device made from semiconductor materials—metals such as germanium or silicon which can act as electrical insulators or conductors, depending on the electrical charges placed upon them. Transistors can be substituted for vacuum tubes in almost all applications involving amplification, rectification, detection, or oscillation. Among the important features transistors possess are: (1) they require no heater current; (2) their small physical size makes them ideal for space-saving applications; (3) their almost unlimited service life.

Transport—The mechanism that moves magnetic tape past the heads.

Tuner—A device that receives radio broadcasts and extracts the audio signal.

Turnover—Frequency above which constant velocity cutting is employed in phonograph recording, and below which a constant amplitude is maintained. In a phono preamp, frequencies below turnover are boosted at the rate of 6 dB per octave to compensate for the cutting process.

Turntable—The circular part of a record-playing device that carries the disc on its circular journey; also the name given a high-quality device for "spinning" records.

Tweeter—A speaker designed to reproduce the high-frequency range of the audio spectrum.

V

Volume—The intensity, or magnitude, of sound.

W

Watt—The unit of measurement of electrical or acoustical power. The rate of work represented by a current of one ampere under a pressure of one volt. Electrical wattage is a measure of the power an amplifier can develop to drive a loudspeaker. Acoustical wattage is a measure of the actual sound a loudspeaker produces in a specific environment. In any amplifier-speaker system, the two figures will differ widely because the low efficiency of speakers requires that they receive high amplifier power to produce satisfactory sound levels over a wide range of frequencies.

Woofer—A speaker designed to reproduce the bass, or low-frequency, portion of the audio-frequency range.

WOW—A form of distortion caused when a tape transport or a phonograph turntable exhibits slow speed variations.

APPENDIX
1
DECIBEL
RELATIONSHIPS

Note that the Power Ratio columns give values which are equal to the number of milliwatts when the reference level is 1 mW, this being also numerically equal to the number of vu for steady sine-wave conditions.

The Voltage Ratio columns also give values which are equal to the number of volts when the reference level is 1 volt.

Interpolation: If it is required to find the power ratio corresponding to 23 dB, or any other value which is not included in the table, the following procedure may be adopted:

1. Take the next lowest multiple of 20 dB (in this case 20 dB), and note the corresponding power ratio (in this example 100).

2. Take the difference between the specified level and the multiple of 20 dB (in this example 23 − 20 = 3 dB) and note the corresponding power ratio (in this example 1.995).

3. Multiply the two power ratios so determined (in this example $100 \times 1.995 = 199.5$).

Decibels Expressed as
Power and Voltage Ratios

Voltage ratio	Power ratio = mW to ref. level 1mW	dB	Voltage ratio	Power ratio = mW to ref. level 1mW
1.0000	1.0000	0	1.000	1.000
.8013	.7943	1	1.122	1.259
.7943	.6310	2	1.259	1.585
.7079	.5012	3	1.413	1.995
.6310	.3981	4	1.585	2.512
.5623	.3162	5	1.778	3.162
.5012	.2512	6	1.995	3.981
.4467	.1995	7	2.239	5.012
.3981	.1585	8	2.512	6.310
.3548	.1259	9	2.818	7.943
.3162	.1000	10	3.162	10.000
.2818	.07943	11	3.548	12.59
.2512	.06310	12	3.981	15.85
.2239	.05012	13	4.467	19.95
.1995	.03981	14	5.012	25.12
.1778	.03162	15	5.623	31.62
.1585	.02512	16	6.310	39.81
.1413	.01995	17	7.079	50.12
.1259	.01585	18	7.943	63.10
.1122	.01259	19	8.913	79.43
.1000	.01	20	10.000	100.00
.056	.00316	25	17.73	316.2
.03162	.001	30	31.62	1,000
.01778	.000316	35	56.23	3,162
.010	.0001	40	100.0	10,000
.0056	.0000316	45	177.8	31,620
.003162	.00001	50	316.2	100,000
.001	.000001	60	1,000	1,000,000
.0003162	.0000001	70	3,162	10,000,000
.0001	.00000001	80	10,000	100,000,000
.00003162	.000000001	90	31,620	1,000,000,000
.00001	.0000000001	100	100,000	10,000,000,000

RESISTOR-CAPACITOR COLOR CODE

Color	Significant figure	Decimal multiplier	Tolerance (%)	Voltage rating*
Black	0	1	—	—
Brown	1	10	1*	100
Red	2	100	2*	200
Orange	3	1,000	3*	300
Yellow	4	10,000	4*	400
Green	5	100,000	5*	500
Blue	6	1,000,000	6*	600
Violet	7	10,000,000	7*	700
Gray	8	100,000,000	8*	800
White	9	1,000,000,000	9*	900
Gold	—	0.1	5	1000
Silver	—	0.01	10	2000
No color	—	—	20	500

*Applies to capacitors only.

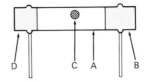

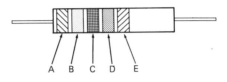

A — First significant figure of resistance in ohms.

B — Second significant figure.

C — Decimal multiplier.

D — Resistance tolerance in percent. If no color is shown the tolerance is ±20 percent.

E — Relative percent change in value per 1000 hours of operation; Brown, 1 percent; Red, 0.1 percent; Orange, 0.01 percent; Yellow, 0.001 percent.

AUDIO-FREQUENCY SPECTRUM

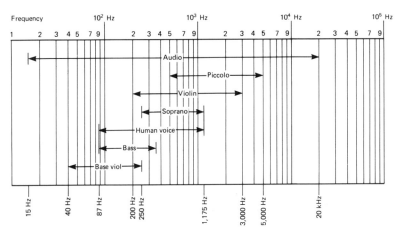

INDEX

I

L

M